Understanding Flight

Understanding Flight

David F. Anderson
Scott Eberhardt

McGraw-Hill

New York San Francisco Washington, D.C. Auckland Bogotá
Caracas Lisbon London Madrid Mexico City Milan
Montreal New Delhi San Juan Singapore
Sydney Tokyo Toronto

Library of Congress Cataloging-in-Publication Data

Anderson, David F.
    Understanding flight / David F. Anderson, Scott Eberhardt.
        p. cm.
    Includes index.
    ISBN 0-07-136377-7
    1. Flight. 2. Aerodynamics. I. Eberhardt, Scott. II. Title.
  TL570 A69 2000
  629.13—dc21

                                   CIP

McGraw-Hill

*A Division of The **McGraw·Hill** Companies*

1 2 3 4 5 6 7 8 9 DOC/DOC 5 4 3 2 1 0

ISBN 0-07-136377-7

The sponsoring editor for this book was Shelley Carr, the editing supervisor was Caroline R. Levine, and the production supervisor was Pamela Pelton. This book was set in Slimbach by Pat Caruso of McGraw-Hill's Professional Book Group composition unit, Hightstown, N.J.

Printed and bound by R. R. Donnelley & Sons Company.

This book is printed on recycled, acid-free paper containing a minimum of 50 percent recycled, de-inked fiber.

McGraw-Hill books are available at special quantity discounts to use as premiums and sales promotions, or for use in corporate training programs. For more information, please write to the Director of Special Sales, McGraw-Hill, Two Penn Plaza, New York, NY 10121-2298. Or contact your local bookstore.

Contents

Introduction

Forget Bernoulli's theorem
WOLFGANG LANGEWIESCHE, STICK AND RUDDER, 1944

There are few physical phenomena so generally studied which are as misunderstood as the phenomenon of flight. Over the years many books have been written about flight and aeronautics (the science of flight). Some books are written for training new aeronautical engineers, some for pilots, and some for aviation enthusiasts. Books written to train engineers often quickly delve into complicated mathematics, which is very useful for those who wish to make detailed calculations. But the necessary formalism is often achieved at the expense of a fundamental understanding of the principles of flight. Books written for pilots and enthusiasts try to explain flight principles but frequently fudge the physics to simplify the explanation.

Unfortunately, the books that do address the principles of flight more often than not propagate long-held myths. We say long-held, but it is interesting to note that if one looks at the description of flight in books written in the 1930s and 1940s one finds essentially the correct explanation. Those discussions focused on the angle of attack and Newton's principles. Somehow between then and now the explanations have gone astray from reality and have become much more complex, nonintuitive, and frequently wrong.

> Sometime between the 1940s and now the explanations of lift have gone astray from reality and have become much more complex, nonintuitive, and frequently wrong.

One common myth is the "principle of equal transit times" which states that the air going around a wing must take the same length of time, whether going over or under, to get to the trailing edge. The argument goes that since the air goes farther over the "hump" on the top of the wing, it has to go faster, and with Bernoulli's principle we have lift. But in reality, equal transit times hold only for a wing *without* lift. Another common misconception is that the shape of the wing is the dominant characteristic in determining lift. Actually, the shape of the wing is one of the least significant features when understanding lift. The principles of lift are the same for a wing flying right side up or in inverted flight.

A shortcoming of many books on the topic of aeronautics is that the information is presented in a very complicated manner, often mistaking mathematics for a physical explanation. This is of little use to one seeking a clear understanding of the basic principles. It is our belief that all fundamental concepts in aeronautics can be presented in simple, physical terms, *without* the use of complicated mathematics. In fact, we believe that if something can only be described in complex mathematical terms it is not really understood. To be able to calculate something is not the same as understanding it.

> To be able to calculate something is not the same as understanding it.

The object of this book is to provide a clear, physical description of lift and of basic aeronautical principles. This approach is useful to one who desires a more intuitive understanding of airplanes and of flight. This book is written for those interested in airplanes in general, and those interested in becoming more proficient pilots. Teachers and students who are looking for a better understanding of flight will find this book useful. Even students of aeronautical engineering will be able to learn from this book, where the physical descriptions presented will supplement the more difficult mathematical descriptions of the profession.

The first chapter, "Basic Concepts," is an introduction to a basic set of terms and concepts. This will give the reader and the authors a common set of tools with which to begin the discussion of flight and aeronautics.

The next chapter, "How Airplanes Fly," is where we get into lift and flight. We believe that this chapter gives the most complete and correct physical description of lift to date. Like many before us, we describe lift using Newton's three laws. But, unlike anyone to our knowledge, we take this description and use it to derive almost all aspects of flight. It allows us to intuitively explain aspects that most aeronautical engineers can only explain mathematically. It will become clear to the reader why one increases the angle of the wing when the airplane slows down or why lift takes less power when the airplane goes faster. It will be obvious why airplanes can have symmetric wings or fly upside down.

The third chapter is "Wings." Here we will explain why wings look the way they do and what tradeoffs take place in their design. When you fly on a commercial jet and see all the changes that are made to the wing on landing, you will have a clear understanding of what is going on and why. A natural follow-on to "Wings" is a chapter on "Stability and Control." Airplane stability is presented and the distinction is made between stable and balanced flight. The concept of fly-by-wire and the role of the computer are also discussed.

Among other things, the chapter on "Propulsion" explains how a jet engine works and why they have gotten so large. Would you believe that there is essentially a propeller in front of the jet engine on that Boeing 777?

The following chapter on "High-Speed Flight" discusses the interesting phenomena associated with flying faster than the speed of sound and why these airplanes look as they do. Chapter 7, "Airplane Performance," discusses such aspects of flight as the climb, cruise, and landings.

The final chapter on "Aerodynamic Testing" addresses wind-tunnel testing and flight testing. The principles of wind tunnels are presented in some detail. Some examples of flight testing as it pertains to the previously introduced concepts are presented.

This work is presented on two levels. The bulk of the material is addressed to the general reader. Here a minimum of experience is assumed. At times it will be desirable to make clarifying comments by insertion of a short topic, which may be somewhat removed from the main train of thought. These insertions are printed on a colored background. They may be skipped over without any loss of continuity or understanding of the main text.

David Anderson
Scott Eberhardt

Understanding Flight

Basic Concepts

A serious discussion of aeronautics requires a basic set of concepts and terminology. Frequently, in the process of learning, the language becomes an unrecognized barrier to the "uninitiated." By starting with some basic terminology and concepts we hope to alleviate some of this.

Airplane Nomenclature

Some readers may be familiar with the language of airplanes and others not. We encourage all to read the following sections of this chapter to ensure that subsequent chapters are easily understood. Those familiar with the major parts of the airplane, the operation of the control surfaces, and the basic operation of an airplane can skip to the section on kinetic energy.

There are over 600,000 licensed pilots in the United States.

The Airplane

Figure 1.1 shows the main components of a high-winged airplane. The *airframe* consists of the *fuselage*, which is the main component of the airplane, the *wings*, and the *empennage*. The empennage (sometimes called the *tail feathers*) is the tail assembly consisting of the *horizontal stabilizer*, the *elevators*, the *vertical stabilizer*, and the *rudder.* The elevators are used to adjust, or control, the pitch (nose up/down attitude) of the airplane. The elevators are connected to the control wheel or stick of the airplane and are moved by the forward and backward motion of the control. On some airplanes the entire horizontal stabilizer is the elevator, as shown in Figure 1.2. This is called a *stabilator.* The rudder is used to make small directional changes and in turns. Two pedals on the floor operate the rudder, used to provide directional control.

Most airplanes have small hinged sections on the trailing edge of the elevators and sometimes on the rudder called *trim tabs* as shown in Figure 1.2. These tabs move in the opposite direction to the control

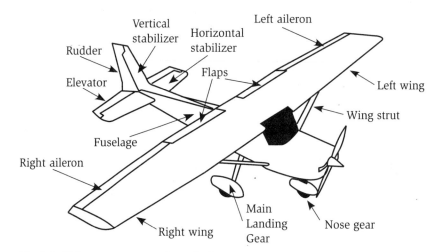

Fig. 1.1. Main components of an airplane.

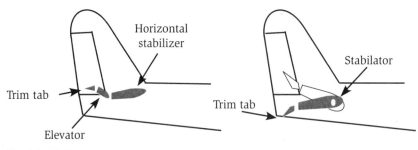

Fig. 1.2. The empennage.

surface. The purpose of the trim tabs is to reduce the necessary force on the control wheel, called a *yoke,* for the pilot to maintain a desired flight attitude.

Most modern airplanes have single wings mounted either above or below the fuselage. Most but not all high-winged airplanes have wings that are supported by *struts.* Struts allow for a lighter wing but at the expense of more drag (resistance to motion through the air).

The movable surfaces on the outer trailing edge of the wings are the *ailerons,* which are used for roll control (rotation around the axis of the fuselage). They are operated by the rotation of the control wheel or by the left-right movement of the stick. The ailerons are coupled so that when one swings up the other swings down. Control surfaces are discussed in detail below.

> A tail dragger is also known as conventional gear because before WWII nose wheels were rare.

The hinged portions on the inboard part of the trailing edge of the wings are the *flaps.* These are used to produce greater lift at low speeds and to provide increased drag on landing. This increased drag helps to reduce the speed of the airplane and to steepen the landing approach angle. Flaps are discussed in detail in the chapter on wings.

Small airplanes have two configurations of landing gear. *Tricycle landing gear* has the *main landing gear* just behind the center of balance of the airplane and a steerable *nose gear* up forward. The *tail dragger* has the main landing gear forward of the center of balance and a small steerable wheel at the tail. The nose gear and the tail wheel are steered with the rudder pedals.

Airfoils and Wings

An *airfoil* is a shape designed to produce lift. As shown in Figure 1.3, an airfoil is the shape seen in a slice of a wing. Besides the wing, propellers and the tail surfaces are also airfoils. Even aeronautical engineers (in discussions) sometimes mistakenly use the terms *wing* and *airfoil* interchangeably. But an airfoil is just the shape seen in a slice of the wing and not a wing itself. For some wings, slices taken at different places along its length will reveal different airfoils.

An airfoil, as shown in Figure 1.3, has a *leading edge* and a *trailing edge.* As detailed in Figure 1.4, a *chord* and a *camber* also characterize an airfoil. The chord is an imaginary straight line connecting the

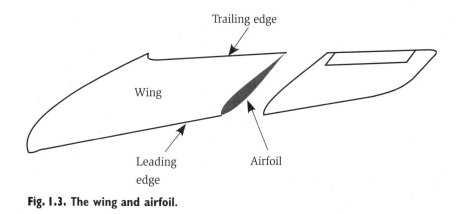

Fig. 1.3. The wing and airfoil.

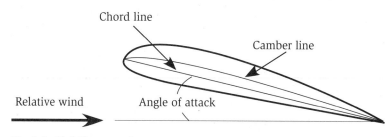

Fig. 1.4. Airfoil nomenclature.

leading edge with the trailing edge. The chord is used for determining the geometric *angle of attack* (discussed below) and for determining the area of a wing.

The *mean camber line* is the line an equal distance from the upper and lower surfaces of the wing. The camber is the curvature of the mean camber line. A wing that has an airfoil with a great deal of curvature in its mean camber line is said to be a *highly cambered* wing. A symmetric airfoil has no camber.

An airfoil with lift also has an angle of attack, as shown in the figure. The *relative wind* is the direction of the wind at some distance from the wing. It is parallel to and opposite to the direction of motion of the wing. The velocity of the relative wind is equal to the speed of the wing. In aeronautics the *geometric angle of attack* is defined as the angle between the *mean chord* of the airfoil and the direction of the relative wind.

A useful measure of a wing is its *aspect ratio*. The aspect ratio is defined as the wing's *span* divided by the average or *mean chord* length. The span is the length of the wing measured from wingtip to wingtip. The mean chord length is the average chord length along the wing. The area of the wing is just the span times the mean chord length. Most wings on small general-aviation airplanes have aspect ratios of about 6 to 8. This means that the wing is 6 to 8 times longer than its average width.

The Boeing B-314 Flying Boats were so large that there was a catwalk in the wings so the mechanics could service the engines in flight.

Axes of Control

An airplane moves in three dimensions called *roll, pitch,* and *yaw,* illustrated in Figure 1.5. Roll is rotation about the *longitudinal axis* that goes down the center of the fuselage. The ailerons control rotation about the roll axis. Pitch is rotation about the lateral axis of rotation, which is an axis parallel to the long dimension of the wings. The elevators control the pitch of the airplane. By controlling the pitch of the airplane, the elevators also control the angle of attack of the wing. To increase the angle of attack, the entire airplane is rotated up. As we will see, this control or the angle of attack is key in the adjustment of

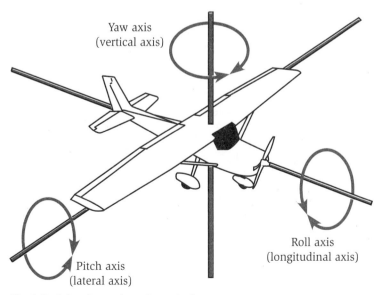

Yaw axis
(vertical axis)

Pitch axis
(lateral axis)

Roll axis
(longitudinal axis)

Fig. 1.5. Axis of rotation of an airplane.

the lift of the wings. Finally, yaw, which is controlled by the rudder, is rotation about the vertical axis, which is a line that goes vertically through the center of the wing. It is important to note that all three axes go through the *center of gravity* (often abbreviated c.g.) of the airplane. The center of gravity is the balance point of the airplane. Or, equivalently, all of the weight of the airplane can be considered to be at that one point.

The Turn

One common misconception by those who are not pilots is that, as with a boat, the rudder is the control used for making a turn. Although very small direction changes can be made with the rudder, the ailerons are use in making turns. The airplane is rolled to an angle in the direction of the desired turn. The lift developed is perpendicular to the top of the wing. In straight-and-level flight, this is straight up. As shown in Figure 1.6, when the airplane rolls to some angle, the direction of lift is now at an angle with part of the lift force used for turning and part used to support the weight of the airplane. In a turn the rudder is only used to make small corrections and *coordinate* the turn.

As shown in the figure, the pilot feels a force equal to the lift but in the opposite direction. Occasionally in the following chapters we refer to the 2*g* turn. A 2*g* turn is a turn where the force felt by the pilot is twice the force of gravity (2*g*) and the force, or *load*, on the wing has been doubled. In aeronautical terms the *load factor*, which is the

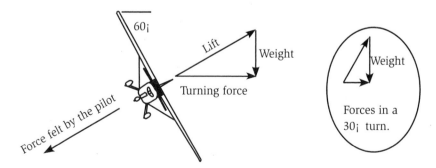

Fig. 1.6. Forces on an airplane in a turn.

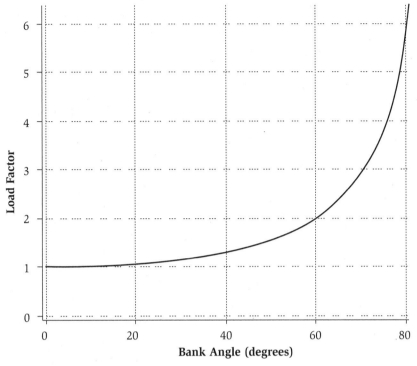

Fig. 1.7. Load factor as a function of bank angle.

load divided by the weight, on the airplane in a *2g* turn is 2. Figure 1.7 shows the load factor as a function of bank angle for any airplane in flight. One thing to understand is that the forces on the pilot (or the load) are only related to the *bank angle,* which is the angle made by the wing and the horizon. In Figure 1.6 the bank angle is 60 degrees. The vertical part of the lift must always be equal to the weight of the airplane if the altitude of the airplane does not change during the turn. This is called a *level turn.* So the steeper the bank angle the greater the lift and thus the greater the force felt by the pilot. The insert in the figure shows the forces of a 30 degree turn for comparison. The weight part of the lift is the same, but the other two forces are less. A *2g* turn is achieved by banking the airplane at an angle of 60 degrees, independent of the speed of the airplane. Turns will be discussed in more detail in the chapter on "Airplane Performance."

Straight-and-level flight

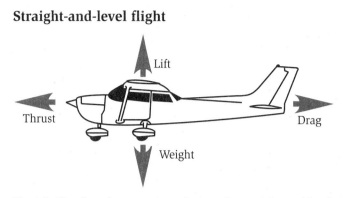

Fig. 1.8. The four forces on an airplane in straight-and-level flight.

The Four Forces

There are four forces associated with the flight of an airplane. These forces, illustrated in Figure 1.8, are lift, weight, thrust, and drag. In *straight-and-level flight* (not changing speed, direction, or altitude) the net lift on the airplane is equal to its weight. We say net lift because, for a conventional airplane design, the horizontal stabilizer pulls down, putting an additional load on the wings. The thrust produced by the engine is equal to the drag, which is caused by air friction and the work done to produce the lift.

The Wright brothers' first flight was less than half the length of a Boeing 777-300.

Mach Number

One important parameter in describing high-speed flight is the *Mach number*. The Mach number is simply the speed of the airplane, or speed of the air, measured in units of the speed of sound. Thus, an airplane traveling at a speed of Mach 2 is going twice the speed of sound. The speed of sound is fundamental for flight because it is the speed of communication between the airplane and the air, and between one part of the air and another. As will be seen in later chapters, as the speed of an airplane approaches Mach 1 there are dramatic changes in its performance. One change that affects performance is that the air ceases to separate before the wing arrives and instead collides energetically with the wing.

The speed of sound is not a constant in air. In particular it changes with air temperature and thus with altitude. As the air temperature decreases with altitude, so does the speed of sound, though not as quickly. At sea level the value of Mach 1 is about 760 mi/h (1220 km/h). The speed of sound decreases with altitude to about 35,000 ft (about 11,000 m), where the value is about 660 mi/h (1060 km/h). The speed of sound then remains essentially constant to an altitude of 80,000 ft (24,000 m). No airplanes fly above this altitude, with the exception of the Space Shuttle on its way back from space.

> The Soviet TU-144 supersonic transport crashed at the Paris Airshow in 1973. It was later found that the pilot was trying to avoid a French Mirage fighter that was trying to take unauthorized pictures of the TU-144 in flight. The TU-144 was planned to compete with the Concorde.

Kinetic Energy

Kinetic energy is the energy of an object because it is moving. The energy difference between a bullet sitting on a table and one flying through the air is the kinetic energy. To be technical, if the bullet had a mass m (say in grams, for example) and were moving at a velocity v, its kinetic energy would be $\frac{1}{2}mv^2$. (There, we have reached the highest level of math complexity that one needs in order to understand flight.)

Because we are going to be discussing the movement of air and the production of propulsion by the acceleration of air or exhaust, it is important that when we say kinetic energy it be understood that we mean the energy due to motion. It is that simple.

Air Pressures

Before we go into understanding flight and airplanes, we should spend just a little time discussing air pressure to remove some common misunderstandings. The following discussion is only strictly true for air moving at speeds below about Mach 0.3 (three-tenths the speed of sound) where the air can be considered incompressible. This will be discussed in greater detail in the next chapter.

Many of us have seen pictures of air passing through a tube that narrows as in Figure 1.9. The figure will often be referred to in text that says something like, "As the area of the tube narrows, the flow velocity must increase. If no other force acts on the fluid, the pressure at point A must be greater than the pressure at point B." This is the Bernoulli relationship that some are familiar with in the explanation of lift in flight. At first the meaning of "the pressure at point A" seems obvious. What is never said in physics books is that the pressure referred to is measured perpendicular to the direction of flow. It is also not said that there are two other pressures associated with the air at point A. One of them has increased and the other has remained the same. The aeronautical engineers understand this concept, but somehow the information has never made it to the aviation community.

As we have said, there are three pressures associated with flowing air. The first is the *total pressure.* This is measured by bringing the flowing air to a stop. In Figure 1.9, this is measured by placing a tube facing into the airflow. The air stops in the tube and the total pressure

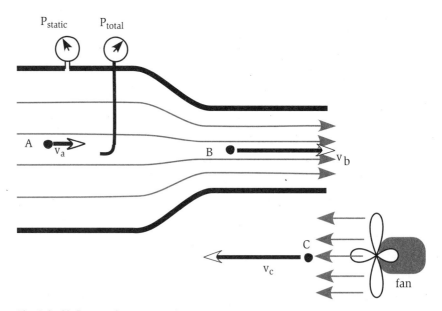

Fig. 1.9. Airflow and pressures.

is measured (P_{total} in the figure). In the situation in the figure, P_{total} is the same at both points A and B. In the language of pilots, this is also known as the *Pitot* pressure, and the figure illustrates a *Pitot tube.*

The second pressure to consider is the *static pressure* (P_{static}), which is measured perpendicular to the airflow through a hole in the wall. This is the pressure most often referred to when the air pressure is discussed in aerodynamics. In the figure the static pressure is higher at point A than at point B.

The third pressure is the *dynamic pressure* ($P_{dynamic}$), which is the pressure due to the motion of the air, and is a pressure parallel to the flow of air. The dynamic pressure is proportional to the kinetic energy in the air. Thus the faster the air goes the higher the dynamic pressure. This may seem a little complicated, so let us try to put it all together.

The total pressure P_{total} is the sum of the static and dynamic pressures (P_{static} and $P_{dynamic}$). We have shown how to measure the total, or Pitot, pressure and the static pressure. How is the dynamic pressure of the air measured? Look at the setup in Figure 1.10. Between the tube that measures the total pressure and the tube that measures the static pressure there is placed a *differential pressure gauge.* That is a gauge that measures the difference in pressure between the two ports, which is the difference between total and static pressure. Since static plus dynamic pressure is equal to total pressure, this difference between total and static pressures is the dynamic pressure.

If no energy is added to the air (by some mechanism such as a propeller) the total pressure remains the same and an increase in dynamic pressure causes a decrease in static pressure. So when one reads that the pressure of air decreases because it is going faster, the pressure referred to is the static pressure. But what if energy is added to the air? In the right-hand corner of Figure 1.9 is the picture of a fan. What has happened to the air pressures at point C? The fan is accelerating the air, and thus has done work on the air. Thus the dynamic pressure has increased. Since the air is not confined, the static pressure is the same as in the surrounding environment and thus has

Alexander Graham Bell was a founding member of the Aerial Experiment Association (AEA), which was organized to try to be the first in flight. It was disbanded in 1908 but had been partially successful with the designs of Glenn Curtiss.

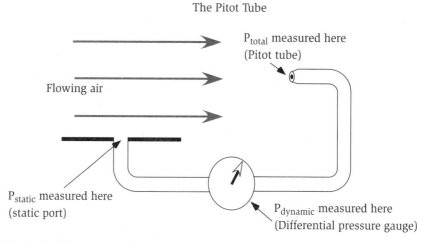

Fig. 1.10. Measurement of pressures with the Pitot tube.

During WWII the length of a belt of 50-caliber machine gun bullets was 27 feet. When a pilot emptied his guns into a single target, he was giving it the "whole nine yards."

not changed. Thus the total pressure has increased. The moral of this story is that when someone refers to air pressure of moving air they are probably referring to static pressure (though they may not know it). It is also wrong to think that just because air is flowing faster the (static) pressure has decreased. This topic is covered in more detail in the Appendix.

The Pitot Tube

As mentioned previously, the tube measuring the total pressure in Figure 1.10 is called the *Pitot tube,* which, along with static pressure, is used for measuring the airspeed of an airplane. Several Pitot tubes can be seen on the front of the large jets at the airport. A single Pitot tube can be seen protruding (or hanging) from the wing of small airplanes. The hole that measures the static pressure of the air is the *static port.* This port is usually on the side of the airplane fuselage near the front, though it is also occasionally placed on the side of the Pitot tube itself. In an airplane there is a gauge that measures the difference between these two devices and is calibrated in speed. This is the *airspeed indicator.* Since the dynamic pressure measured is related to the dynamic pressure, there is also a correction made by the pilot for the reduced density of the air at altitudes above sea level.

In 1908, a French woman named Therese Peltier became the first woman pilot.

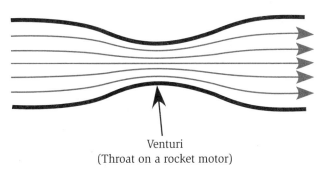

Venturi
(Throat on a rocket motor)

Fig. 1.11. The venturi or throat.

Venturi and Throat

Take a look at Figure 1.11. Here the tube has a constriction in it. In a tube this constriction is often referred to as a *venturi.* If the constriction is in a rocket motor, it is called a *throat.* Off and on in the book we will find occasion to refer to both the venturi and the throat.

Other Books

The purpose of this book is to give the reader a physically intuitive understanding of flight and of aerodynamics, without the use of complicated mathematics. But, at some point one may want to go beyond a merely physical understanding to a level of understanding that does require more mathematics. The authors have found two books very useful in the writing of this book.

The first book is *Aerodynamics for Naval Aviators* by H. H. Hurt, Jr., published by Direction of Commander, Naval Air Systems Command, United States Navy. This book is a wealth of information and insight for those who want to go into greater detail in understanding modern flight.

The second book is *Introduction to Flight,* by John D. Anderson, published by McGraw-Hill. This book is certainly a modern classic in the field of aerodynamics textbooks. *Introduction to Flight* is a serious college textbook and is geared to those with a good understanding of mathematics.

Both are excellent references for those seriously interested in airplanes and aerodynamics. Of the two, *Aerodynamics for Naval*

The world's worst airplane accident was in Tenerife, Canary Islands, on March 27, 1977. Two Boeing 747s collided on the ground, killing 561 people.

Aviators is a better reference for the person interested in a good aeronautical reference without the need of calculus.

Wrapping It Up

The brief discussion of nomenclature, controls, turns, and pressure should give you the necessary tool to understand the rest of the book. In the next chapter, How Airplanes Fly, we present a physical description of flight. After flight is understood, we discuss the airplane itself.

CHAPTER 2

How Airplanes Fly

As mentioned in the introduction of this book, a great deal of false concepts and "mythology" have built up around the principles of flight. In this chapter we explain with simple logical discussions the physical phenomena of lift and address some of the errors in the present explanations. Armed with an understanding of lift, we take it further and give you an intuitive understanding of flight in a much broader sense. We start by looking at three descriptions of lift.

The Popular Description of Lift

Most of us have been taught what we will call the "popular description of lift," which fixates on the shape of a wing. The key point of the popular description of lift is that the air accelerates over the top of the wing. Because of the Bernoulli effect, which relates the speed of the air to the static pressure, a reduced static pressure is produced above the wing, creating lift. The missing piece in the description is an understanding of the cause of the acceleration of the air over the top of the wing. A clever person contributed this piece with the introduction of the "principle of equal transit times," which states that the air that separates at the leading edge of the wing must rejoin at the

15

Daniel Bernoulli did not derive the "Bernoulli equation." His friend Leonid Euler did.

trailing edge. Since the wing has a hump on the top, the air going over the top travels farther. Thus it must go faster to rejoin at the trailing edge. The description is complete.

This is a tidy explanation and it is easy to understand. But one way to judge an explanation is to see how general it is. Here one starts to encounter some troubles. If this description gives us a true understanding of lift, how do airplanes fly inverted? How do symmetric wings (the same shape on the top and the bottom) fly? How does a wing flying at a constant speed adjust for changes in load, such as in a steep turn or as fuel is consumed? One is given more questions than answers by the popular description of lift.

One might also ask if the numbers calculated by the popular description really work. Let us look at an example. Take a Cessna 172, which is a popular, high-winged, four-seat airplane. The wings must lift 2300 lb (1045 kg) at its maximum flying weight. The path length for the air over the top of the wing is only about 1.5 percent greater than the length under the wing. Using the popular description of lift, the wing would develop only about 2 percent of the needed lift at 65 mi/h (104 km/h), which is "slow flight" for this airplane. In fact, the calculations say that the minimum speed for this wing to develop sufficient lift is over 400 mi/h (640 km/h). If one works the problem the other way and asks what the difference in path length would have to be for the popular description to account for lift in slow flight, the answer would be 50 percent. The thickness of the wing would be almost the same as the chord length.

Though enthusiastically taught, there is clearly something seriously wrong with the popular description of lift. The first thing that is wrong is that the principle of equal transit times is not true for a wing with lift. It is true only for a wing without lift. Figure 2.1 shows a computer simulation of the airflow around a wing. Periodically simulated smoke has been introduced to show the changes in the speed of the airflow.

The principle of equal transit times is not true for a wing with lift.

The first thing to notice is that the air going over the top of the wing reaches the trailing edge before the air that goes under the wing. In fact, the air that passed under the wing has a somewhat retarded velocity compared to the velocity of air some distance from the wing. Without the principle of equal

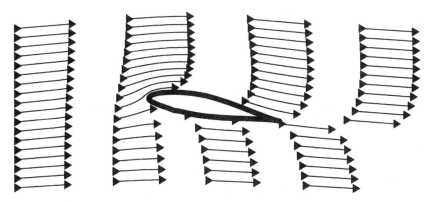

Fig. 2.1. Simulated smoke flowing around an airfoil.

transit times, the popular description of lift loses its explanation for acceleration of the air over the top of the wing.

In explaining the physics of the lift on the wing, we will reveal other problems with this description. But first we would like to introduce two other descriptions of lift. These are the "mathematical description of lift" and the "physical description of lift" The latter is the description followed in this book.

The Mathematical Description of Lift

Aeronautical engineers use the mathematical description of lift, a name coined for the sake of this discussion. It uses complex mathematics and computer simulations, and is a powerful design tool. Typically, the velocities of the air around the wing are generated with a computer program. Then, using Bernoulli's equation, the pressures and lift forces are accurately calculated. Since this description often calculates the lift of a wing from the acceleration of the air, this description has quite a bit in common with the popular description of lift.

The aeronautical engineers know that the principle of equal transit times is not true. They often use a mathematical concept called *circulation* to calculate the acceleration of the air over the wing. Circulation is a measure of the apparent rotation of the air around the wing. In the mathematical description of lift circulation is used to discuss why the air goes more slowly below the wing and faster over the top. While useful for calculations of lift, circulation is not something a pilot needs to consider while flying an airplane. The concept of circulation is useful, however, for the understanding of upwash and of ground effect, and is discussed in greater detail later in this chapter.

Like the popular description of lift, the classical theories in the mathematical description of lift do not illustrate the fact that lift requires the action of a great deal of air, much more than you might imagine. We will not spend much time on the mathematical description of lift, though it will come up from time to time.

The mathematical description of lift is a general term for the analysis tools of classical aerodynamics and computational aerodynamics. If the objective is to accurately compute the aerodynamics of a wing, these are the tools to use, though the description is mathematical and not physical. This is a point lost on many of its proponents. Fortunately, the physical description of lift, presented here, does not require complicated mathematics.

The Wright brothers knew nothing of Bernoulli's principle or any mathematical aerodynamic theories. They just watched how birds fly.

The Physical Description of Lift

The physical description of lift is based primarily on Newton's three laws and a phenomenon called the *Coanda effect*. This description is uniquely useful for understanding the phenomena associated with flight. It is useful for an accurate understanding of the relationships in flight, such as how power increases with load or how the stall speed increases with altitude. It is also a useful tool for making rough estimates ("back-of-the-envelope calculations") of lift. The physical description of lift is also of great use to a pilot who needs an intuitive understanding of how to fly the airplane.

In this description, lift is recognized as a reaction force, that is, wings develop lift by diverting air down. One knows that a propeller produces thrust by blowing air back and that a helicopter develops lift by blowing air down. Propellers and helicopter rotors are simply rotating wings. Thus the concept of a wing diverting air down to produce lift should not be difficult to accept. As we will see, the low pressure that is formed above the wing accelerates the air down, with almost all of the lift of a well-designed wing coming from the diversion of air from above the wing.

Wings develop lift by diverting air down.

One should be careful not to form the image in their mind of the air striking the bottom of the wing and being deflected down. This is a fairly common misconception, and was also made by Sir Isaac Newton himself. Since Newton was not familiar with the details of airflow over a wing, he thought that the air was diverted down by its impact with the bottom of birds' wings. It is true that there is some lift due to the diversion of air by the bottom of the wing, but the majority of the lift is due to the top of the wing.

The physical description of lift is very powerful. It allows one to have an intuitive understanding of such diverse phenomena as the dependence of lift on the angle of attack of the wing, inverted flight, ground effect, high-speed stalls, and much more. As stated before, this description of lift is that it is useful to the pilot in flight. This cannot be said for the popular description of lift.

Newton's Three Laws

The most powerful tools for understanding flight are Newton's three laws of motion. They are simple to understand and universal in application. They apply to the flight of the lowly mosquito and the motion of the galaxies. Let us look at these laws, not quite in order.

A statement of Newton's first law is:

> A body at rest will remain at rest, and a body in motion will continue in straight-line motion unless subjected to an external applied force.

In the context of flight this means that if an object (such as a mass of air), initially motionless, starts to move, there has been force acting on it. Likewise, if a flow of air bends (such as over a wing), there also must be a force acting on it.

Newton's third law can be stated:

> For every action there is an equal and opposite reaction.

This is fairly straightforward. When you sit in a chair, you put a force on the chair and the chair puts an equal and opposite force on you. Another example is seen in the case of a bending flow of air over a wing. The bending of the air requires a force from Newton's first law. By Newton's third law, the air must be putting an equal and opposite force on whatever is bending it, in this case the wing.

Newton's second law is a little more difficult but also more useful in understanding many phenomena associated with flight. The most common form of the second law is

$$F = ma,\ \text{or force equals mass times acceleration}$$

The law in this form gives the force necessary to accelerate an object of a certain mass. For this work we use an alternative form of Newton's second law that can be applied to a jet engine, a rocket, or the lift on a wing. The alternate form of Newton's second law can be stated:

The force (or thrust) of a rocket is equal to the amount of gas expelled per time, times the velocity of that gas.

The amount of gas per time might be in units such as pound mass per second (lbm/s) or kilograms per second (kg/s). The velocity of that gas might be in units such as feet per second (ft/s) or meters per second (m/s). This form of Newton's second law says that if one knows the rate at which gas is expelled from a rocket motor, and the speed of that gas, the thrust of the rocket motor can be easily calculated. To double the thrust, one can double the amount of gas expelled per second, double the velocity of the gas, or a combination of the two.

Let us now look at the airflow around a wing with Newton's laws in mind. In Figure 2.2 we see the airflow around the wing as many of us have been shown at one time or another. Notice that the air approaches the wing, splits, and reforms behind the wing going in the initial direction. This wing has no lift. There is no net *action* on the air and thus there is no lift, the *reaction* on the wing. If the wing has no net effect on the air, the air cannot have any net effect on the wing. Now, look at another picture of air flowing around a wing (Figure 2.3). First the air comes from below the wing. This is the *upwash,* which will be explained later. The air splits around the wing and leaves the wing at a slight downward angle. This downward-traveling air is the *downwash* and as we will see is the source of lift on a wing. In this figure there has been a net change in the air after passing over the wing. Thus there has been a force acting on the air and a reaction force on the wing. There is lift.

If the wing has no net effect on the air, the air cannot have any net effect on the wing and there is no lift.

The final question to be answered before we can go on to really understand lift is why does the air bend around the wing? The answer is in an interesting phenomenon called the *Coanda effect.*

The Coanda Effect

The Coanda effect has to do with the bending of fluids around an object. For the forces and pressures associated with low-speed flight air is considered not only a fluid but an incompressible fluid. This

Fig. 2.2. Based on Newton's laws, this airfoil has no lift.

Upwash

Downwash

Fig. 2.3. The airflow around a real airfoil looks like this.

means that the volume of a mass of air remains constant and that flows of air do not separate from each other to form voids (gaps). For the moment let us consider the Coanda effect with water. This effect can be demonstrated in a simple way. Run a small stream of water from a faucet and bring a horizontal water glass over to it until it just touches the water, as in Figure 2.4. As in the figure, the water will wrap partway around the glass. From Newton's first law we know that for the flow of water to bend there must be a force on it. The force is in the direction of the bend. From Newton's third law we know that there must be an equal and opposite force acting on the glass. The same phenomenon causes forces between the airflow around a wing and the wing. So why do fluids tend to bend around a solid object?

The answer is *viscosity,* that characteristic that makes a fluid thick and makes it stick to a surface. When a moving fluid comes into contact with a solid object, some of it sticks to the surface. A small distance from the surface the fluid has a small velocity with respect to the object. As in Figure 2.5, the farther one looks from the surface, the faster the fluid is flowing, until it eventually comes to the speed of the uniform flow some distance away from the object. The transition layer between the surface and the fluid at the uniform flow is called the *boundary layer,* which is discussed in the next chapter.

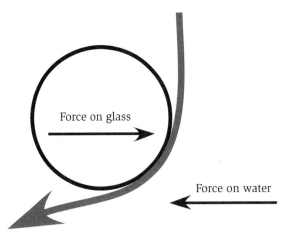

Force on glass

Force on water

Fig. 2.4. The Coanda effect.

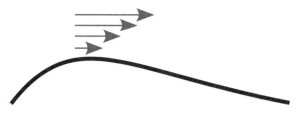

Fig. 2.5. The variation of the speed of a fluid near an object.

The differences in speed in adjacent layers cause *shear forces,* which cause the flow of the fluid to want to bend in the direction of the slower layer. This causes the fluid to try to wrap around the object. The characteristic of fluids to have zero velocity at the surface of an object explains why one is not able to hose dust off of a car. At the surface of the car the water has no velocity and thus puts little or no force on the dust particles.

> The characteristic of fluids to have zero velocity at the surface of an object explains why one is not able to hose dust off of a car.

Viscosity and Lift

For those with some familiarity with aerodynamics, there may be some confusion with the connection of viscosity and lift. Many simulations in aerodynamics are done with zero viscosity or, more accurately, in the limit of zero viscosity. Viscosity is introduced implicitly with the Kutta-Joukowski condition, which requires that the air come smoothly off at the trailing edge of the wing. So, in reality these "zero viscosity" calculations reintroduce viscosity via the Kutta-Joukowski condition. In a fluid without viscosity, such as superfluid helium, a wing could not fly.

Also, in most mathematical descriptions of lift the boundary layer is considered so small that it is ignored. Many erroneously claim that ignoring the boundary layer is equivalent to having zero viscosity. This is incorrect because viscosity is implicit in the condition that air follows the curvature of the wing.

Lift on a Wing

We now have the tools to understand why a wing has lift. In brief, the air bends around the wing because of the Coanda effect. Newton's first law says that the bending of the air requires a force on the air, and Newton's third law says that there is an equal and opposite force on the wing. All this is true. But there is a little more to it than that. First, let us look at the air bending around the wing in Figure 2.6. To bend the air requires a force. As indicated by the gray arrows, the direction of the force on the air is perpendicular to the bend in the air. The magnitude of the force is proportional to the "tightness" of the bend. The tighter the air bends the greater the force on it. The forces on the wing, as shown by the black arrows in the figure, have the same magnitude as the forces on the air but in the opposite direction. When the air bends around the surface of the wing it tries to separate from the airflow above it. But since it has a strong reluctance to form voids, the attempt to separate lowers the pressure and bends the adjacent streamlines above. The lowering of the pressure propagates out at the speed of sound, causing a great deal of air to bend around the wing. This is the source of the lowered pressure above the wing and the production of the downwash behind the wing. Figure 2.7 is a good example of the effect of downwash behind an airplane. In the picture the jet has flown above

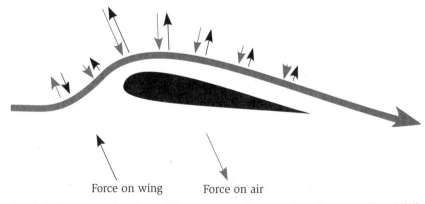

Force on wing   Force on air

Fig. 2.6. Forces on the air and the corresponding reaction forces on the airfoil.

Fig. 2.7. A jet flying over fog demonstrates downwash. (*Photographer Paul Bowen; photo courtesy of Cessna Aircraft Co.*)

the fog, not through it. The hole caused by the descending downwash is clearly visible.

It should be noted that the speed of the uniform flow over the top of the wing is faster than the *free-stream velocity,* which is the velocity of the undisturbed air a great distance from the wing. The bending of the air causes the reduction in pressure above the wing. This reduction in pressure causes acceleration of the air via the Bernoulli effect. Below the wing the uniform flow is slower than the free-stream velocity, because the pressure is slightly increased.

Look again at Figure 2.6, while paying attention to the black arrows representing the forces on the wing. There are two points to notice. The first is that most of the lift is on the forward part of the wing. In fact, half of the total lift on a wing is typically produced in the first one-fourth of the chord length. The second thing to notice is that the arrows on the leading part of the wing are tilted forward. Thus the force of lift is pulling the wing along as well as lifting it. This would be nice if it were the entire story. Unfortunately, the horizontal forces on the trailing part of the wing compensate the horizontal forces on the leading part of the wing. Thus the net result is that the forces produce lift only on the wing. We are ignoring skin friction for the moment.

This is a good time to look back at the popular description of lift. It attributes lift to the acceleration of the air over the wing, causing a reduction in pressure. As stated, the bending of the air causes the reduction in pressure above the wing. This reduced pressure causes the acceleration of the air via the Bernoulli effect. The acceleration of air over the top of a wing is the result of the lowered pressure and not the cause of the lowered pressure.

The acceleration of air over the top of a wing is the result of the lowered pressure and not the cause of the lowered pressure.

The pressure difference across the wing is the cause of lift. But the lowering of the pressure above the wing is the result of the production of the downwash. As we will see, it is the adjustment of this downwash in direction and magnitude which allows the wing to adjust for varying loads and speeds.

Downwash

From Newton's second law one can state the relationship between the lift on a wing and its downwash.

> The lift of a wing is proportional to the *amount* of air diverted per time, times the *vertical velocity* of that air.

Similar to the rocket, one can increase the lift of a wing by increasing the amount of air diverted, the vertical velocity of that air, or a combination of the two. The concept of the vertical velocity of the downwash may seem a little foreign at first. We are all used to thinking of the airflow across a wing as seen by the pilot, or as seen in a wind tunnel. In this "rest frame" the wing is stationary and the air is moving. But, what does the world look like in the rest frame where the air is initially standing still and the wing is moving? Picture yourself on top of a mountain. Now suppose that just as a passing airplane is opposite you, you could take a picture of all the velocities of the air. What would you see? You might be surprised.

Though we go into this in greater detail in a later section, the first thing you would notice is that the air behind the wing is going almost straight down when seen from the ground. (Because of friction with the wing it in fact has a slight forward direction.) The plausibility of this statement is fairly easy to demonstrate. Turn on a small household fan and examine the tightness of the column of air. If the air were coming off the trailing edges of the fan blades (which are legitimate wings) other than perpendicular to the direction of the blades' motion, the air would form a cone rather than a column. This can also be seen in the picture of a helicopter hovering above the water (Figure 2.8). The pattern on the water is the same size as the rotor blades. It is fortunate that nature works this way. If the air behind the propeller of an airplane came off as a cone rather than a tight column, propellers would be a much less efficient means of propulsion. The air expanding out on one side of the airflow would just compensate for the air expanding out on the other side and would not contribute to the thrust.

There are currently over 13,000 airports in the United States.

The wing develops lift by transferring momentum to the air. Momentum is mass times velocity. In straight-and-level flight the momentum is transferred toward the earth. This momentum eventually strikes the earth. If an airplane were to fly over a very large scale, the scale would weigh the airplane. This should not be confused with the (wrong) concept that the earth somehow supports the airplane. It does not. Lift on a wing is very much like shooting a bullet at a tree. The lift is like the recoil that the shooter feels, whether the bullet hits the tree or not. If the bullet hits the tree, the tree experiences the event but has nothing to do with the recoil of the gun.

The air behind the wing is going almost straight down when seen from the ground.

Fig. 2.8. A helicopter pushes air down. (*Photo courtesy of the U.S. Air Force.*)

DOES THE EARTH SUPPORT THE AIRPLANE?

Some insist that since the airplane exerts a force on the earth, in straight-and-level flight, the earth is somehow holding the airplane up. This is definitely not the case. The lift on the wing has nothing to do with the presence of the surface of the earth. Examining two simple examples can show this.

The first example is to consider the thrust of a propeller, which is just a rotating wing. It certainly does not develop its thrust because of the presence of the surface of the earth. Neither could the presence of the earth provide the horizontal component of lift in a steep bank.

The second example to consider is the flight of the Concorde. It cruises at Mach 2, 55,000 ft (16,000 m) above the earth. The pressure information of the jet cannot be communicated to the earth and back faster than the speed of sound. By the time the earth knows the Concorde is there, it is long gone.

The Adjustment of Lift

We have said that the lift of a wing is proportional to the amount of air diverted per time, times the vertical velocity of that air. And we have also stated that in the rest frame where the air is initially at rest and the wing is moving, the air is moving almost straight down after the wing passes. Let us look at this in more detail. Figure 2.9*a* shows how the downwash as seen by the pilot and the observer on the ground relates to the speed of the airplane and the angle of attack of the wing. The arrow marked "Speed" represents the speed and direction of the wing through the air. The arrow marked "Downwash" is the speed and direction of the downwash as seen by the pilot, and the arrow marked "Vv" is the speed and direction of the downwash as seen by an observer on the ground. Vv is the vertical velocity of the downwash and represents the component that produces lift. In this figure the letter α indicates the angle of the wing's downwash with

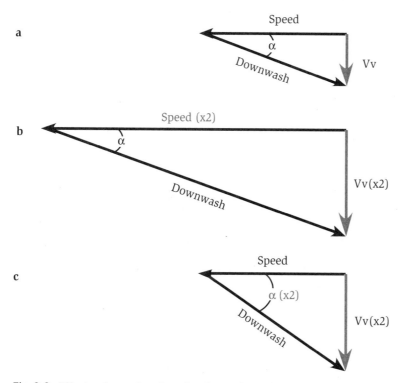

Fig. 2.9. Effects of speed and angle of attack on downwash.

respect to the relative wind, which is proportional to the angle of attack of the wing.

So what would happen if the speed of the wing were to double and the angle of attack remained the same? This is shown in Figure 2.9b. As you can see, the vertical velocity (Vv) has doubled. As we will soon see, the *amount of air diverted has also doubled.* Since both the amount of air diverted and the vertical velocity of the air have doubled with the doubling of the speed, the lift of the wing has gone up by a factor of 4.

In Figure 2.9c the wing has been kept at the original speed but the angle of attack has been doubled. Again the vertical velocity of the air has doubled and the lift of the wing has doubled. What these figures show is that the vertical velocity of the air is proportional to both the speed and the angle of attack of the wing. Increase either and you increase the lift of the wing.

> The vertical velocity of the air is proportional to both the speed and the angle of attack of the wing.

Angle of Attack

Now let us look in more detail at the angle of attack of the wing. In aeronautics the *geometric angle of attack* is defined as the angle between the *mean chord* of the wing (a line drawn between the leading edge of the wing and the trailing edge) and the direction of the relative wind. For our discussion we are going to use the *effective angle of attack.* The effective angle of attack is measured from the orientation where the wing has zero lift. The difference between the geometric angle of attack used in aeronautics and the effective angle of attack used here should be emphasized to prevent potential confusion by the reader. Figure 2.10 shows the orientation of a cambered wing with zero *geometric* angle of attack, and the same wing with a zero *effective* angle of attack. A cambered wing at zero geometric angle of attack has lift since there is a net diversion of the air down. By definition the same wing at zero effective angle of attack has no lift and there is no net diversion of the air. In the case of a symmetric wing, the geometric and effective angles of attack are of course the same.

For any wing, from that of a Boeing 777 to a wing in inverted flight, an orientation into the relative wind can be found where there is zero lift. As the wing is rotated from this position, the change in angle is the

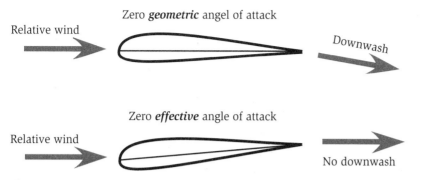

Fig. 2.10. Definition of geometric and effective angles of attack.

effective angle of attack. Now if one starts with the wing at zero degrees and rotates it both up and down while measuring the lift, the response will be similar to that shown on the graph in Figure 2.11. This is an extremely important result. It shows that the lift of a wing is proportional to the *effective* angle of attack. This is true for all wings: those of a modern jet, wings in inverted flight, a barn door, or a paper airplane. A similar graph would be found for lift as a function of the geometric angle of attack, but the line would be different for different wings, with only the symmetric wing passing through the origin.

As can be seen in Figure 2.11, the relationship between lift and the angle of attack breaks down at the *critical angle*. At this angle the forces become so strong that the air begins to separate from the wing and the wing loses lift while experiencing an increase in drag, a retarding force. At the critical angle the wing is entering a *stall*. The subject of stalls will be covered in detail in the next chapter.

> The lift of a wing is proportional to the angle of attack. This is true for all wings, from a modern jet to a barn door.

The Wing as a Scoop for Air

Newton's second law tells us that the lift of a wing is proportional to the amount of air diverted down times the vertical velocity of that air. We have seen that the vertical velocity of the air is proportional to the speed of the wing and to the angle of attack of the wing. We have yet to discuss how the amount of air is regulated. For this we would like

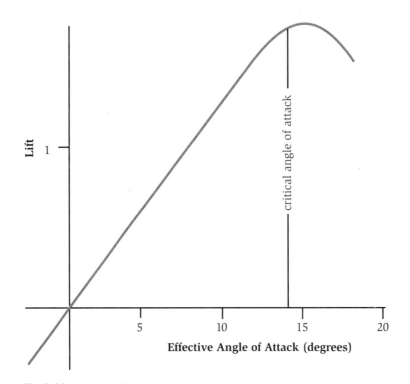

Fig. 2.11. Lift as a function of angle of attack.

to adopt a *visualization tool* of looking at the wing as a scoop (Figure 2.12) that intercepts a certain amount of air and diverts it to the angle of the downwash. To be more accurate, the air diverted at the bottom of the scoop has a vertical velocity defined by the speed of the wing and its angle of attack. As one goes farther from the wing, the vertical velocity decreases until at the top of the scoop the air has a very small vertical velocity. For wings of typical airplanes it is a good approximation to say that the area of the scoop is proportional to the area of the wing. The shape of the scoop is approximately elliptical for all wings, as shown in the figure. Since the lift of the wing is proportional to the amount of air diverted by the wing, which is proportional to the area of the wing, the lift of a wing is proportional to its area.

If one were to move such a scoop through the air, how much air would it divert? Certainly, if the scoop were to divert a certain amount of air at one speed, it would divert twice as much air at twice the

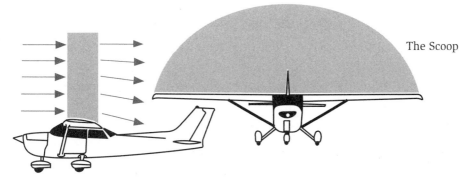

The Scoop

Fig. 2.12. The scoop as a visualization tool.

speed. But what if the scoop were taken to a higher altitude where the air has a lower density? If the air were half as dense the scoop would divert half as much air for a given speed. Thus, the amount of air intercepted by a wing is proportional to its area, the speed of the airplane, and the density of the air.

The amount of air diverted by a wing is proportional to its area, the speed of the airplane, and the density of the air.

Temperature and humidity also affect air density. An increase in temperature causes the air to expand. An increase in temperature from 32°F (0°C) to 95°F (35°C) causes a reduction in air density of 10 percent. Humidity reduces the density of the air because water vapor is almost 40 percent less dense than the air it displaces.

For the normal changes in angle of attack experienced in flight we can make the statement that the amount of air diverted by the wing is not dependent on the angle of attack of the wing. The amount of air diverted also does not depend on the load on the wing, which affects the angle of attack.

We said that the scoop is only a visualization aid, but it is a very useful one in understanding how much air a wing diverts. It also gives correct aerodynamic relationships such as how the lift goes with speed, air density, angle of attack, and power. The relationships that we derive using the scoop image are exact and not approximations.

Let us do a back-of-the-envelope calculation to see how much air a wing might divert. Take, for example, a Cessna 172 that weighs about 2300 lb (1045 kg). Traveling at a speed of 140 mi/h (220 km/h) and assuming an effective angle of attack

Chuck Yeager became an ace in a day.

of 5 degrees, we get a vertical velocity for the air of about 11.5 mi/h (18 km/h) right at the wing. If we assume that the *average* vertical velocity of the air diverted is half that value, we calculate from Newton's second law that the amount of air diverted is on the order of 5 tons/s. Thus, a Cessna 172 at cruise is diverting about five times its own weight in air per second to produce lift. Think how much air is diverted by a 250-ton Boeing 777.

> A Cessna 172 at cruise is diverting about five times its own weight in air per second to produce lift.

So how big is the scoop? If for simplicity we take the scoop to be rectangular for the calculation, with a length equal to the wingspan of 36 ft (14.6 m), we get a height of about 18 ft (7.3 m). This is a lot of air. One should remember that the density of air at sea level is about 2 lb/yd^3 (about 1 kg/m^3). As implied by the shape of the scoop in Figure 2.12, the lift is greatest at the root of the wing tapering to the wingtip. Thus the air is diverted from considerably farther above the root of the wing than the 18 ft calculated here.

The popular description of lift, and to a lesser extent the mathematical description of lift, discusses the effect of the wing on the air only very near its surface. This gives the false impression that lift is a very local effect involving a small amount of air. Our back-of-the-envelope calculation of the amount and extent of the air involved in the lift of a wing shows that this is not true. A great deal of air is involved in the production of lift.

This large amount of diverted air causes the lower wing of a biplane to interfere with the lift of the upper wing. The air diverted by the lower wing reduces the air pressure on the bottom of the upper wing. This reduces the lift and efficiency of the upper wing. Thus, many biplanes have the upper wing somewhat forward of the lower wing, or at least the root of the upper wing is moved forward to reduce this interference.

> A Boeing 747-400 flying at maximum range is 45 percent fuel, by weight, at takeoff.

Putting It All Together

You now know that the lift of a wing is proportional to the amount of air diverted times the vertical velocity of that air. The amount of air diverted by the wing is proportional to the speed of the wing and the density of the air. The vertical velocity of the downwash is proportional to the angle of attack and the speed of the wing. With this knowledge we are in a position to understand the adjustment of lift in flight.

As our first example, let us look at what happens if the load on the wing is increased by the airplane going into a 2*g* turn. In such a turn the load on the wings has doubled. If we assume that the speed of the airplane is kept constant, the vertical velocity of the downwash must be doubled to compensate for the increased load. Doubling the angle of attack does this.

During peak WWII production, Boeing built 17 B-17s a day in a single plant.

Now what happens when an airplane flying straight and level doubles its speed? If the pilot were to maintain the same angle of attack, both the amount of air and the vertical velocity of the downwash would double. Thus the lift would go up by a factor of 4. Since the weight of the airplane has not changed, the increased lift would cause the airplane to increase altitude rapidly. So to maintain a constant lift, the angle of attack must be decreased to decrease the vertical velocity. If the speed of an airplane were increased by 10 percent, the amount of air diverted by the scoop would increase by 10 percent. Thus the angle of attack would have to be decreased to give a 10 percent reduction in the vertical velocity of the downwash. The lift of the wing would remain constant.

As our last example, let us consider the case of an airplane going to a higher altitude. The density of the air decreases and so, for the same speed, the amount of air diverted has decreased. To maintain a constant lift, the angle of attack is increased to compensate for this reduction in the diverted air. If the density of the air is reduced by 10 percent the vertical velocity of the downwash is increased by the same 10 percent to compensate. This is accomplished by increasing the angle of attack by that amount.

We now understand how the airplane adjusts the lift for varying load, speed, and altitude. The next step in understanding the flight of an airplane is the subject of power and drag. We will start with a look at power.

"Penny-planes" are rubber-band-powered model airplanes that can weigh no more than a penny.

Power

One of the most important concepts for understanding flight is that of the power requirements. In aeronautics textbooks the discussion of drag, which is a force against the motion of the airplane, would come first and power would be given little consideration. That may be appropriate for the design of an airplane, but it is less useful for the understanding of its operation.

Power is the rate at which work is done. The power associated with flight also relates to the demand placed on the engine and the limitations on airplane performance. We will consider two types of power requirements. The first is *induced power,* which is the power associated with the production of lift. It is equal to the rate at which energy is transferred to the air to produce lift. So when you see the word *induced* with respect to flight, think of lift. The second power requirement we need to consider is *parasitic power.* This is the power associated with the impact of the air with the moving airplane. The *total power* is simply the sum of the induced and parasitic powers.

Induced Power

Let us first look at the induced power requirement of flight. The wing develops lift by accelerating air down. Before the wing came by, the air was standing still. After the wing passes, the air has a downward velocity, and thus it has been given kinetic energy. As stated in Chapter 1, if one fires a bullet with a mass m and a velocity v, the energy given to the bullet is simply $\frac{1}{2}mv^2$. Since the induced power is the rate at which energy is transferred to the air, it is proportional to the *amount of diverted air* times the *vertical velocity squared* of that air.

During an evacuation of a Chinese village in 1942, 60 passengers were loaded on a DC-3 designed for 30. When the plane landed, there were 68 passengers plus the crew of four. Eight were stowaways. One of the passengers was Brigadier General James Doolittle, returning from his famous raid on Tokyo.

(Remember that in the rest frame of the observer on the ground the direction of the downwash is down.) But since the lift of a wing is proportional to the amount of air diverted times the vertical velocity of that air, we can make a simplification. *The induced power associated with flight is proportional to the lift of the wing times the vertical velocity of the air.* Now let us look at the dependence of induced power on the speed of the airplane.

We know that if the speed of an airplane were to double, the amount of air diverted would also double. So the angle of attack must be adjusted to give half the vertical velocity to the air, maintaining a constant lift. The lift is constant and the

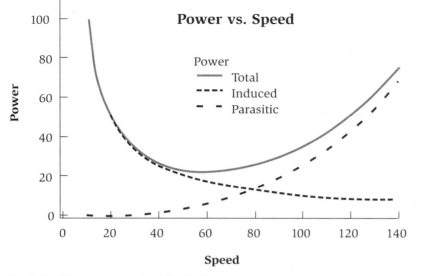

Fig. 2.13. The power required for flight.

vertical velocity of the downwash has been halved. Thus, the induced power has been halved. From this we can see that *the induced power varies as 1/speed for a constant load.* The induced power is shown as a function of speed by the dotted line in Figure 2.13. This shows that the more slowly the airplane flies the greater the power requirement to maintain lift. As the airplane slows in flight, more and more power must be added until finally the airplane is flying at full power with the nose high in the air. What is happening is that as the airplane's speed is reduced, more and more energy must be given to less and less air to provide the necessary lift.

The induced power varies as 1/speed for a constant load.

Parasitic Power

Parasitic power is associated with the energy lost by the airplane to collisions with the air. It is proportional to the average energy that the airplane transfers to an air molecule on colliding times the rate of collisions. As with the energy given to the bullet above, the energy lost to the air molecules is proportional to the airplane's speed squared. The rate of collisions is simply proportional to the speed of the airplane. The faster the airplane goes the higher the rate of collisions. So we have a speed squared due to the energy given to each molecule and a single speed term due to the collision rate. This yields the result that *the parasitic power varies as the speed cubed.* The parasitic power as a function of speed is also graphed in Figure 2.13 by the dashed line.

The fact that the parasitic power goes as the airplane's speed cubed has an important consequence for the performance of an airplane at its cruise speed, where it is limited by the parasitic power. In order for an airplane to double its cruise speed, it would have to increase the size of its engine by eight times! So when an airplane owner upgrades to a larger engine, there is an improvement in the rate of climb and turn of the airplane but only a modest increase in cruise speed. To substantially increase the speed of the airplane, the parasitic power must be decreased. Such design features as retractable landing gear, smaller fuselage cross sections, and an improved wing design accomplish this.

The Power Curve

As stated above, the total power is the sum of the induced and parasitic powers. The solid line in Figure 2.13 shows the total power as a function of speed. At low speed the power requirements of the airplane are dominated by the induced power which goes as 1/speed. At cruise speeds the performance is limited by the parasitic power which goes as speed cubed. This graph of total power as a function of speed is known as the *power curve.* Flying at slow speeds where the total power requirement increases with decreasing speed is what pilots refer to as flying the *backside of the power curve.*

One might ask how an increase in altitude would affect the power curve. This is illustrated in Figure 2.14, which shows the power curves for altitudes of 3000 ft and 12,000 ft (about 900 m and 3600 m). With an increase in altitude, there is a decrease in air density. Thus, the wing diverts less air and the angle of attack must be increased in order to maintain lift. As stated before, as the density of the air is reduced, the angle of attack, and the vertical velocity of the downwash, must be increased to compensate. Thus, the induced power would be increased. A 10 percent reduction in air translates to approximately a

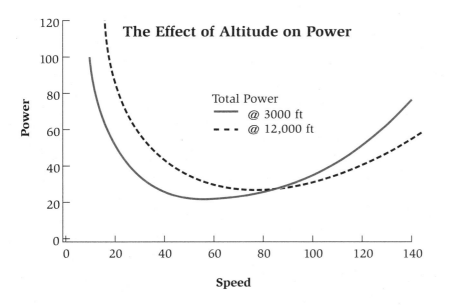

Fig. 2.14. Total power required for flight at two altitudes.

10 percent increase in induced power. An airplane flying on the backside of the power curve would require more power and fly with a greater angle of attack when going to a higher altitude.

The situation is the opposite for the parasitic power. A reduction in air density translates to a reduction in the number of collisions with the air, and thus there is a reduction in the parasitic power. An airplane at cruise speed where parasitic power dominates finds it more economical to fly at a higher altitude. Usually flying at a higher altitude does not translate into flying at a higher speed because nonturbocharged engines experience a reduction in power that is similar to the reduction in atmospheric pressure. That is, if the atmospheric pressure is 65 percent that of sea level, the maximum power of the engine is also approximately 65 percent of its sea-level performance.

> Before 1900, Langley's law, found experimentally by Samuel Langley, said that total power required for flight decreases with speed. It has been shown that all of his experiments were performed on the backside of the power curve.

The Effect of Load on Induced Power

Now let us examine the effect of load on induced power. First, remember that the induced power associated with flight is proportional to the *lift* of the wings times the *vertical velocity* of the down-wash. Now if we were to double the load, maintaining the same speed, we would have to double the vertical velocity of the air to provide the necessary lift. Both the load and the vertical velocity of the air have been doubled and the induced power has gone up by a factor of 4. Thus, *the induced power increases as the load squared.* It is easy to see why the weight of an airplane and its cargo is so important. Figure 2.15 shows the data for the relative fuel consumption of a heavy commercial jet as a function of weight. These measurements were made at a fixed speed. From the data one can estimate that at a gross weight of 500,000 lb (227,000 kg) and a speed of Mach 0.6 about 40 percent of the power consumption is induced power and 60 percent is parasitic power. In reality the airplane would cruise at a speed of around Mach 0.8, where the induced power would be lower and the parasitic power consumption would be higher. Unfortunately, at that more realistic speed the details of the power consumption become more complicated and it is more difficult to separate the parasitic and induced powers from the data.

The induced power increases as the load squared.

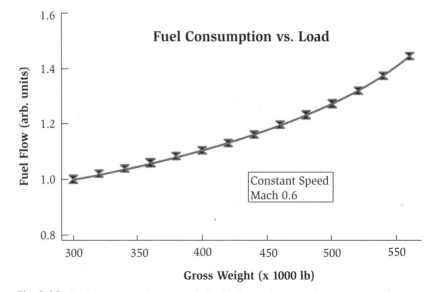

Fig. 2.15. Fuel consumption vs. weight for large jet at a constant speed.

Drag

So far, we have discussed power at length with only brief references to the topic of drag. With an understanding of power we are in a position to understand drag, which is part of the pilot's culture and vocabulary. Drag is a force that resists the motion of the airplane. Clearly, a low-drag airplane will fly faster than a high-drag airplane. It will also require less power to fly the same speed as the high-drag airplane. So, what is the relationship between power, drag, and speed?

Power is the rate at which work is done. In mathematical terms it is also a force times a velocity. Drag is a force and is simply equal to power/speed. We already know the dependence of induced and parasitic powers on speed. By dividing power by speed, we have the dependence of drag on speed. Since induced power varies as 1/speed, *induced drag* varies as 1/speed squared. Parasitic power varies as speed cubed, so *parasitic drag* varies as speed squared. Figure 2.16 shows the dependence of induced, parasitic, and total drag on the speed of the airplane.

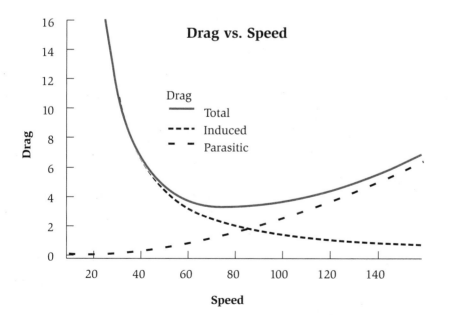

Fig. 2.16. Drag as a function of speed.

The Wright Flyer did not move to the Smithsonian until 1948. The Wright brothers were snubbed when the Smithsonian decided to display Samuel Langley's Aerodrome as the "first airplane capable of flight." Langley had been the secretary of the Smithsonian.

In the section preceding this one we saw that the induced power increases as load squared. Since drag is just power divided by speed, induced drag also increases as load squared. Anything understood about power can be easily converted to a similar understanding of drag by simply dividing by speed.

We have said that drag is part of a pilot's culture and vocabulary. That is true. But most of the time when the term is used, the person really means power. Let us look at an example to illustrate. Take the case of a pilot flying a small plane with retractable landing gear. If full power is applied in straight-and-level flight, the airplane accelerates to some speed and goes no faster. A pilot might well say that the airplane's speed is limited by the drag.

Let us pretend that an airplane had two meters, one that measured total drag and another that measured the total power for flight. We will then record both values for the airplane at its top speed. The pilot lowers the landing gear and flaps, leaving the engine at full power. There is now a substantial increase in the drag and power required. This of course slows the airplane down. We would find that the airplane slowed down to the previous total power requirement and now the total drag is higher than before. The pilot would have to reduce power to bring the total drag down to its previous value. Its top speed was not determined by the total drag but by the total power (drag times speed). So, when pilots say drag they usually mean power.

The utility of the concept of power over drag for the pilot is fairly easy to understand. Power requirements relate simply to the demands on the engine. Drag is a force that must be related to the airplane's speed in order to understand the related power requirement to overcome it. A drag of a certain value at one speed is only half the power drain of a drag of the same value at twice the speed. In the end, the power available from the engine is what counts.

The Wing's Efficiency for Lift

We have seen that the induced power requirement of a wing varies as 1/speed for a fixed load and as load squared for a fixed speed. But one may wonder how the design of the wing affects the induced power requirements. In other words, what is the wing's efficiency for lift?

Efficiency for lift has to do with the amount of induced power it takes to produce a certain lift. The lower the induced power needed the greater the efficiency. The most obvious way to improve the efficiency of a wing is to increase the amount of air diverted by the wing. If more air is diverted, the vertical velocity of the air is reduced for the same lift and so is the induced power. This is accomplished by increasing the size of the wing.

Approximately 600 million people fly on domestic commercial routes per year. That's over 1.6 million a day!

Another important contribution to the induced power requirements of a wing is the additional loading due to the upwash. In brief, the upwash puts an additional load on the wing on the order of 20 percent of the lift for a general-aviation airplane. This additional load decreases with increasing aspect ratio. This is discussed in the section on ground effect.

In consideration of the total efficiency of a wing, the parasitic power (or parasitic drag) must also be considered. The parasitic power of a wing is proportional to its area. So for cruise speeds where parasitic power dominates there is a limit to how much the area of the wing can be increased to reduce the induced power. There are additional problems with increasing the wing's area, particularly by making it longer. The first is that large wings are heavy and increase the weight of the airplane. The second is that long wings are not as structurally strong as shorter wings. If two wings have the same area, the longer one will be the most efficient due to reduced upwash loading, though it will be weaker. Gliders operate at speeds where induced power dominates. Thus the high-performance gliders have long wings. Figure 2.17 shows a glider with a 60:1 glide ratio. This means that in still air, the glider travels 60 ft horizontally for every foot it descends to provide power. It is interesting to note that sea birds, which must fly long distances without landing, also have high aspect ratio wings for optimum efficiency.

Sea birds, which must fly long distances without landing, also have high aspect ratio wings for optimum efficiency.

Most fast airplanes have shorter wings. An exception is the U-2 spy plane, which flew at 460 mi/h (740 km/h) above 55,000 ft (16,700 m). The U-2 had long wings because of the extreme altitudes at which it

Fig. 2.17. A high-performance glider. (*Photo courtesy of Motorbuch Verlag, Stuttgart, Germany, from Cross-Country Soaring.*)

operated. Because of the low density of the air, the induced power is significant at its cruise speed.

The Physics of Efficiency

Let us now look at the physics associated with the efficiency of lift. This will help us understand such things as why helicopters are less efficient than fixed-wing aircraft and what affects the efficiency of propulsion systems (discussed in Chapter 4). The first point to consider is that the force producing the lift is proportional to the momentum (mv, where m is the mass of the air and v is the vertical velocity) that is transferred to the air per time. Thus, either the acceleration of more mass or the acceleration of the air to a higher velocity will increase the lift. The lift, of course, is the desired end product.

The induced power consumed is proportional to the amount of kinetic energy ($\frac{1}{2}mv^2$) that is given to the air. For a given lift the energy consumed must be minimized. That is, we want to produce the desired lift for the least induced power. Thus one must make m as large as possible and reduce v to as small as possible. For maximum lift efficiency one must accelerate a large amount of air at as low a velocity as possible. This gives the desired lift with the least energy given to the air.

> For maximum lift efficiency one must accelerate a large amount of air at as low a velocity as possible.

Because the rotors on a helicopter are quite small for the weight of the aircraft, they must accelerate a relatively small amount of air to a high velocity (high kinetic energy) to produce the needed lift. This is inefficient. A similar argument can be made to understand why an airplane cannot "hang" on its propeller. Although the engine is producing sufficient power to lift the airplane with the wings, the propeller accelerates too little air at too high a velocity to produce the necessary lift on its own. One airplane that can hang on its propellers is the Bell-Boeing V-22 Osprey tilt-rotor aircraft shown in Figure 2.18. The extreme propellers, or *proprotors,* divert a great deal of air, allowing the engines to produce enough thrust to lift the craft vertically off the ground before they are rotated forward for horizontal flight.

Lift Requires Power

Sometimes one encounters the misconception that a wing requires very little power to produce lift and that there is no net downwash behind the wing. This misconception is easy to understand. In many computations in aerodynamics the calculations are done with two-dimensional airfoils. These are in fact infinite wings. This is done because an infinite wing is much easier to calculate than one of finite span. The efficiency of a wing increases with the span of the wing, since the amount of air diverted increases with area. Thus, an infinite wing diverts an infinite amount of air at zero velocity to produce lift, and thus is infinitely efficient. The net vertical velocity of the downwash is zero. Therefore, the infinite wing requires no power to produce lift. Of course, this is not the situation for a real three-dimensional wing.

Fig. 2.18. The V-22 Osprey. (*Used with the permission of the Boeing Management Company.*)

Wing Vortices

The downwash behind the wing is sometimes called the *downwash sheet*. This downwash sheet has a curl in it, producing the *wing vortex*. Near the tip of the wing the wing vortex curls very tightly, creating the *wingtip vortex*. The wingtip vortex initially contains a small amount of the energy of the wing vortex, but because it is often so visible, it is the only part of the wing vortex with which most people are familiar. Eventually, the entire wing vortex curls into a single trailing vortex on each side. To understand why the wing vortex curls, we must first consider the lift distribution of the wing.

In our discussion of the scoop, we illustrated in Figure 2.12 that the amount of air diverted by a wing is a maximum near the root and decreases to zero at the wingtip. The height of the scoop at any point along the wing represents the load and the momentum transferred at that point. The load on a wing is nicely illustrated in Figure 2.19, which shows the condensation on top of the wing of a fighter aircraft during a high-*g* maneuver. The lowered pressure above the wing reduces the air

Fig. 2.19. Illustration of the lift distribution on the wings of an F-14A. (*Photo courtesy of NASA.*)

temperature, causing condensation in the form of a kind of fog. This fog displays the variation in load along the wing. One can see that the load is greatest near the root of the wing and it tapers to zero at the wingtip. The large fuselage causes the lift to decrease along the centerline of the fighter. The lift must go to zero at the wingtip because the relatively high pressure under the wing communicates around the wingtip with the lowered pressure above the wing.

The greater the load on a section of wing, the greater the amount of air that section diverts and the farther above and behind the wing the effect of the wing is felt. Let us look at the downwash sheet behind the wing. At the trailing edge of a wing the downwash is roughly constant in velocity. But as one goes farther back from the wing, the sections with greater load will have a higher vertical velocity, as shown in Figure 2.20. So, after the wing passes, the load along the wing expresses itself in the change in velocity of the downwash, with the higher load having the highest velocity.

In the discussion of the Coanda effect, we saw that a difference in velocity in adjacent layers of air caused the air to bend toward the more slowly moving air. Likewise the difference in velocity of the downwash sheet causes the air to bend, wrapping from higher load to

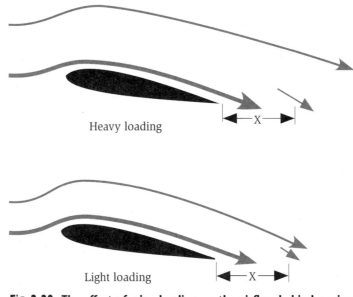

Heavy loading

X

Light loading

X

Fig. 2.20. The effect of wing loading on the airflow behind a wing.

lower load. The tightness of the bend reflects the rate of change in the load along the wing. At the end of the wing the lift goes to zero very rapidly and there is some airflow around the wingtip. This causes the tightest curl in the wing vortex, creating the wingtip vortex. The details of the wing vortex can be clearly seen in Figure 2.7 of a jet flying over fog. The center of the trough formed by the downwash sheet has smooth sides terminating in the tight curl of the wingtip vortex.

> At the end of the wing the lift goes to zero very rapidly, causing the tightest curl in the wing vortex and creating the wingtip vortex.

The wingtip is usually the place on the wing with the greatest change in lift and thus produces the most visible vortex. This is not always at the wingtip. Figure 2.21 shows a landing airplane producing *flap vortices*. In this example the change in lift is greatest at the outer edge of the wing's flap.

Circulation

As mentioned before, circulation is a measure of the rotation of the air around the wing, when seen from the rest frame where the air is initially standing still and the wing is moving. Circulation has been mis-

Fig. 2.21. Wing flap vortices. (*Photographer, Jan-Olov Newborg.*)

takenly employed by some as a *driving* mechanism to accelerate the air over the top of the wing and thus account for the reduction of pressure causing lift. Let us go back to the rest frame of an observer on a mountaintop who is able to take a picture of the directions of air movement around a wing as it passes. What would the picture look like? It would look something like Figure 2.22. Remember when studying this figure that it is a snapshot of a moving wing and that the arrows represent the velocities in the air at one moment in time. The air is not moving around the wing but is shifting in a circular pattern. The arrow marked "1" will become the arrow marked "2" in a moment, and so on. If one

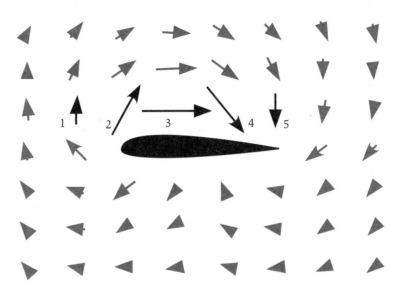

Fig. 2.22. Airflow around a wing as seen from an observer watching on the ground.

adds the speed of the direction and speed of the relative wind (as seen by the wing) to each of the arrows in the figure, the familiar streamlines with upwash and downwash are produced.

So, why is this happening? First, we have to bear in mind that air is considered an incompressible fluid for these discussions. That means that it cannot change its volume and that there is a resistance to the formation of voids. Now the air has been accelerated over the top of the wing by the reduction in pressure. This draws air from in front of the wing and expels if back and down behind the wing. This air must be compensated for, so the air shifts around the wing to fill in. This is similar to the circulation of the water around a canoe paddle. This circulation around the wing is no more the driving force for the lift on the wing than is the circulation in the water that drives the paddle. It is true that if one is able to determine the circulation around a wing the lift of the wing can be calculated. Lift and circulation are proportional to each other.

The most obvious result of the circulation around the wing is that the air approaches from below the wing. This is the

> Eddie Rickenbacker, the U.S. "ace of aces" in WWI, was a famous auto racer before the war and owned the Indianapolis Speedway after the war. He also owned and built up Eastern Airlines.

cause of upwash. When an airplane is traveling at Mach 1 or faster, information cannot communicate forward and the wing has no upwash. But at the speed common for small airplanes, upwash is quite pronounced. If Figure 2.22 were expanded so more of the air could be seen, one would see that the air is going almost straight down far behind the wing, and that the effect of the wing extends far above the wing.

Something to notice in the simulations of circulation is that there is very little action below the wing. Most of the work is being done above the wing. This is why the bottom side of military aircraft can be so cluttered with munitions and fuel tanks, as in the F-16 fighter shown in Figure 2.23. These obstructions cause an increase in parasitic power but do little to affect the efficiency of the wing. The top of the wing is a different story. Obstructions above the wing interfere with lift. This explains why struts are common on the bottom of wings but are historically rare on the top of the wing.

Flight of Insects

Statements have been made that classical aerodynamic theory proves that insects cannot fly. In all cases that we are familiar with, the flight of insects is expressed in terms of circulation. Circulation is a model developed for large aircraft that does not apply to small insects, by blowing air down. Insects obey the same laws of physics as airplanes. Insects produce lift just like airplanes, by blowing air down. When you have a chance, observe a bumblebee feeding on flowers. You will see that when it flies over a leaf on the plant, the leaf is depressed just as if it had landed on it. It clearly is producing lift the same way an airplane does.

Ground Effect

The concept of circulation is necessary for the understanding of *ground effect*. This effect is the increase in efficiency of a wing as it

Insects produce lift just like airplanes, by blowing air down.

Fig. 2.23. **Armament goes on the bottom of the wing of an F-16.** *(Photo courtesy of the U.S. Air Force.)*

comes to within about a wing's length of the ground. The effect increases with the reduction in the distance to the ground. A low-wing airplane will experience a reduction in the induced drag of as much as 50 percent just before touchdown. This reduction in drag just above a surface is used by large birds, which can often be seen flying just above the surface of the water. Pilots taking off from deep-grass or soft runways also use ground effect. The pilot is able to lift the airplane off the soft surface at a speed too slow to maintain flight out of ground effect. This reduces the resistance on the wheels and allows the airplane to accelerate to a higher speed before climbing out of ground effect.

What is the cause of this reduction in drag? Many have the misconception that it is air piling up between the wing and the ground, and that the airplane is flying something like a hovercraft. To understand ground effect we first examine the air flowing over a wing in greater detail. Notice in Figure 2.24 that the air bends up from its horizontal flow to form the upwash. Newton's first law says that there must be a force acting on the air to bend it. Since the air is bent up, the force must be up as shown by the arrow. Newton's third law says that there is an equal and opposite force on the wing which is down. The result is that the upwash increases the load on the wing. In straight-and-level flight the bending air over the top of the wing must now lift the weight of the airplane plus the additional load caused by

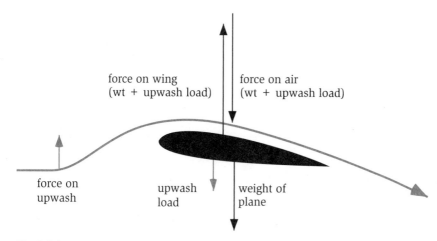

Fig. 2.24. Wing out of ground effect.

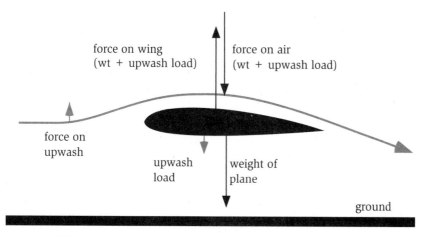

Fig. 2.25. Wing in ground effect.

the upwash. To compensate for this increased load, the wing must fly at a greater angle of attack, and thus a greater induced power. As the wing approaches the ground, the circulation below the wing is inhibited. As shown in Figure 2.25, there is a reduction in the upwash and in the additional loading on the wing caused by the upwash. To compensate, the angle of attack is reduced and so is the induced power. The wing becomes more efficient.

Until now we have not considered the additional loading caused by upwash in the description of flight. The total load on the wing is equal to the lift of the wing plus the upwash loading. This additional load due to upwash is equal to $(2/AR)*\text{lift}$, where AR is the wing's aspect ratio (span/mean chord). Most small airplanes have aspect ratios of 7 to 8. An airplane with an aspect ratio of 7 will experience approximately a 20 percent reduction of wing loading in ground effect. Since induced power is proportional to the load squared, this corresponds to almost a 40 percent reduction in induced power. The glider in Figure 2.17 gets a substantial reduction in wing loading due to upwash from its very long wings.

It is reasonable to ask if this additional loading due to upwash changes the physical description of flight. The answer is no. The increased loading is proportional to the lift. Let us look at an airplane going through a 2g turn at a constant speed. Before the turn the wing had a certain load due to the weight of the airplane and to the upwash. In a 2g turn the apparent weight of the airplane is doubled, and the load due to upwash has also

doubled. Thus the load on the wing has doubled. Since the angle of attack must also be doubled to maintain altitude, nothing has changed in our description. All the descriptions above remain the same. It is just that the load on the wing is higher than the weight of the airplane, but still proportional to it.

Wrapping It Up

The key thing to remember about lift is that it is a reaction force caused by the diversion of air down. The lift of a wing is proportional to the *amount* of air diverted times the *vertical velocity* of that air. For a given wing the amount of air diverted is proportional to the *speed* of the wing and the *density* of the air. The vertical velocity of the downwash is proportional to the *angle of attack* and the *speed* of the wing. The induced power is proportional to 1/*speed* and to *load squared*. The parasitic power of an airplane is proportional to *speed cubed*. With these basic concepts the phenomena of flight can be easily understood.

In this chapter we have downplayed the importance of the shape of the wing. It is obvious to anyone who has been around airplanes that wings are in fact very complicated structures with a great deal of engineering involved. In the following chapter the wing is considered in detail.

CHAPTER 3

Wings

In the last chapter you were introduced to the physical description of flight. There the role of the shape of the wing to lift was downplayed and modern airplane wings and barn doors were treated as equals. In this chapter you will learn how the concepts used to understand flight can help to understand the design of a wing. There are many factors that go into the design of a wing. Should the wing be swept back and tapered? What airfoil should be used? What high-lift devices should be added to improve takeoff and landing performance? These are just some of the questions that must be answered when designing a wing.

Besides aerodynamic considerations, the wing designer must consider other tradeoffs such as structural weight and cost. Some aerodynamically sound principles have fallen prey to the realities of construction costs or structural weight. Understanding these tradeoffs is more a function of experience than of formal training. Hopefully, by reading this chapter you will gain an appreciation for the decisions that must be made.

Airfoil Selection

Before a wing can be designed, a wing section, or airfoil, must be selected. As stated in Chapter 1, the airfoil is a slice of a wing as viewed from the side (see Figure 1.3). In the previous chapter it was emphasized that lift is primarily a function of angle of attack, with little dependence on the airfoil shape. So, why will not almost any wing section do? Here we will discuss some of the specific airfoil design characteristics that are used and how they affect performance. Characteristics that must be considered in selecting a wing include lift at cruise angle of attack, drag, stall characteristics, laminar flow, and room for internal structures.

Wing Incidence and Camber

The lowest drag for the fuselage will be achieved when the fuselage is aligned with the relative wind. A symmetric wing would have to be attached at an "incident angle" so that the fuselage is at a zero angle to the wind while the wing is at some positive angle of attack. When a child builds a simple hobby-shop glider with flat balsa wings, the wings are attached at an angle to the fuselage. Alternatively, a wing with sufficient camber can be selected such that when the leading and trailing edges are aligned at cruise the lift balances the weight. This is when the geometric angle of attack is zero (Figure 2.10). This configuration gives lift with a low parasitic drag and parasitic power. Typically, cambered wings are used and mounted to the fuselage with a small incident angle for cruise conditions.

It might seem that a highly cambered airfoil would give the best results. The appropriate amount of camber will depend on the application. An acrobatic airplane, which spends much of the time upside down, will have a symmetric airfoil section. Modern aircraft frequently have positive camber over most of the airfoil but a reversed, or negative, camber toward the trailing edge. This reversed camber can reduce drag at high speeds. So, selecting a correct airfoil for a particular application involves complex choices and can become more of an art than a science.

The projected lifetime of a commercial airplane design is 70 years.

Wing Thickness

Wing thickness is another design consideration. A thick wing can result in a large wake, resulting in high parasitic drag, even at zero angle of attack. The airflow around a thick wing may separate, causing *form drag,* a type of parasitic drag.

A thicker wing will have a structural advantage, since the wing's structure can be contained inside the wing itself. The thin wings used by early airplanes required external bracing and thus higher parasitic drag. Before aerodynamic drag was understood toward the end of WWI, emphasis was placed on thin wings, with many wires and struts to give the wing strength. Biplanes were used most frequently because they made for a nice boxlike structure. There is a natural competition between the aerodynamicist who wishes to have a thin wing and the structural engineer who wants the wing to look internally like a nice fat box.

The Fokker D-VII was the first airplane to exploit thick wings. As a result it was the only airplane specifically mentioned in the Treaty of Versailles. The Germans were required to hand over every airplane to the Allies.

Leading Edge

Another design characteristic to be considered in designing a wing is the shape of the leading edge. A wing with a sharp leading edge will go abruptly into a stall. A blunt leading edge will have a much less abrupt stall entry, with the stall starting at the trailing edge with the separation point progressing forward as the stall becomes deeper. Figure 3.1 shows the early stall development of a wing with a well-rounded leading edge as well as the stall of a wing with a sharp leading edge. The wing with the sharp leading edge goes directly into a full stall. Another way of looking at stall entry is shown in Figure 3.2, which shows the lift as a function of effective angle of attack for both wings. It is clear that the sharp wing goes from maximum lift to full stall with a very small change in angle of attack.

Airplanes that are designed to operate at lower speeds or to be used as trainers have fairly round leading edges. Fast jet fighters have sharper leading edges. A difficult problem for pilots transitioning from lower-speed trainers to high-speed fighters is appreciating the more abrupt stall. Many new pilots have been "surprised" when their high-speed airplane stalled without warning. A classic use of a sharp leading edge wing section is on the F-104 Star Fighter, shown in Figure 3.3. The leading edge is so sharp it has been known to cut the hands of mechanics who inadvertently rub against it.

STOL (short takeoff and landing) aircraft have very fat, round leading edges. This will hurt the airplanes' top cruise speed, but top

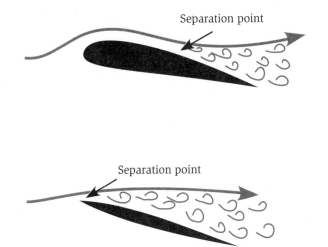

Fig. 3.1. Flow separating from two different airfoils.

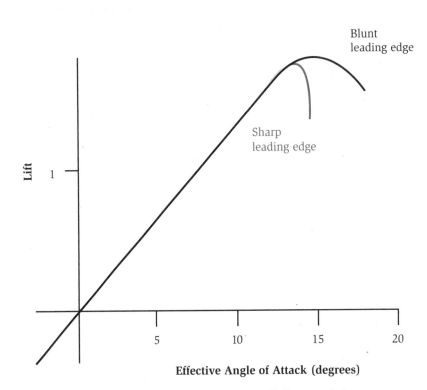

Fig. 3.2. The effect of leading edge shape on stall characteristics.

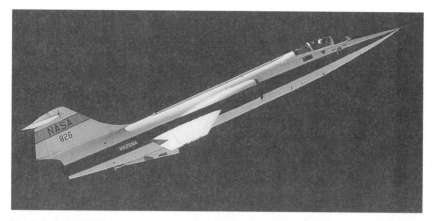

Fig. 3.3. The F-104 Star Fighter with a thin sharp wing. (*Photo courtesy of NASA.*)

Fig. 3.4. A STOL airplane with a fat leading-edge wing. (*Photo courtesy of Czech Aircraft.*)

cruise speed is usually not the primary goal of STOL aircraft. Figure 3.4 shows the wing of a typical STOL aircraft. It is much fatter than for a typical general-aviation airplane and has a well-rounded leading edge.

Wing Planforms

Along with selecting the right airfoil, or wing section, the designer must also select the right *planform*. The planform is the contour of the wing as viewed from above. How big should the wing be? Should it have a high or a low aspect ratio? Should the wing be swept? Should it be tapered? In this section we examine how to choose the basic planform of the wing.

Wing Loading

The first design parameter to determine is the wing loading. This is the ratio of wing area to weight of the airplane, measured in lb/ft^2 or kg/m^2. Many basic performance parameters are determined as a function of wing loading, which is discussed in Chapter 7, "Airplane Performance." Wing loading will determine cruise performance, take-off and landing distances, and power requirements. Typical wing loadings for various aircraft are shown in Table 3.1. Note that light trainers have low wing loading while commercial transports and military aircraft have a wing loading as much as 10 times greater.

There are many tradeoffs to be considered with wing loading. One consideration is that the higher the wing loading the higher the stall speed. That is why trainers have lower wing loading. But low wing loading also limits top cruising speed. A higher wing loading is desired for faster airplanes. Also, airplanes with higher wing loading are less susceptible to clear air turbulence. The inertia of the airplane against the gusts of turbulence makes the airplane harder to blow around. Details of these tradeoffs are discussed in Chapter 7.

Aspect Ratio

The aspect ratio of the wing is the span (measured from tip to tip) divided by the mean chord. In Chapter 2 the advantage of high aspect ratio (long narrow wings) was discussed for low-speed aircraft. High aspect ratio wings are more efficient, because of the reduced loading due to upwash. A high aspect ratio wing can use a smaller engine and needs less takeoff and landing distance than a low aspect ratio wing. However, there are many low-speed aircraft with low aspect ratio wings. Why would a designer select a low aspect ratio when it is known that the wing will suffer induced power and drag penalties?

Insects have a wing loading of 0.01 to 0.5 lb/ft^2 (1 to 10 N/m^2). Birds have a wing loading of 0.1 to 1.0 lb/ft^2 (10 to 100 N/m^2). Airplanes have a wing loading from 10 to 200 lb/ft^2 (1000 to 20,000 N/m^2).

Table 3.1 Typical wing loading (lb/ft^2)

Type	Wing Loading (lb/ft^2)	Wing Loading (kg/m^2)
Sailplane	6	30
General aviation (single engine)	17	85
General aviation (twin engine)	26	130
Jet fighter	70	350
Jet transport	120	600

The reason may be structural considerations. A long slender wing requires more material. A lower aspect ratio wing may save enough on structural weight to offset the reduced lift efficiency, since induced power goes as the load squared. Since weight is probably the singularly most important design criterion, the final wing design must take weight into account. A lighter wing means a lighter airplane and this allows the designer to use a smaller wing area. The smaller wing weighs even less, so there is a cascading positive effect of designing a lighter wing. However, at low speeds the decision to use a low aspect ratio wing will result in higher induced power, so a larger engine might be necessary.

Another reason to choose a low aspect ratio wing, despite its inefficiency, may be for aerobatics. A high aspect ratio wing will not roll as quickly as a low aspect ratio wing. In aerobatics the difference in roll rate can be the edge in a competition. The same is true for fighter aircraft. In close combat it is desirable to have a more maneuverable plane than the adversary.

In contrast, airplanes that are designed for high-altitude flight, such as the U-2 reconnaissance airplane, have very high aspect ratio wings, as can be seen in Figure 3.5. At high altitudes the amount of air available to divert is very small so airplanes must fly at a high angle of attack. In Chapter 2 we discussed that high angles of attack mean greater induced power. High aspect ratio wings reduce induced power and are the wing of choice for high altitude. A notable exception is the SR-71 Blackbird shown in Figure 3.6. But that airplane has other performance criteria that set it apart from other high-

A wandering albatross has an aspect ratio of 20, equivalent to a standard-class sailplane, while the highly maneuverable humming bird has an aspect ratio of less than 7. The YF-22 stealth fighter (Figure 3.36) has an aspect ratio of 2.4.

Fig. 3.5. A U-2 with a high aspect ratio wing. (*Photo courtesy of the U.S. Air Force.*)

Fig. 3.6. The SR-71B spy plane. (*Photo courtesy of NASA.*)

altitude airplanes. This exception will be discussed in Chapter 6, "High-Speed Flight."

Sweep

Most modern aircraft use swept wings, as shown in the picture of an X-5 with variable-sweep wings (Figure 3.7). The primary motivation behind swept wings is to reduce drag at higher cruise speeds. It was discovered in Germany in the late 1930s that at high speeds the parasitic drag of the wing was related to the angle the air makes with the wing's leading edge. Thus, by sweeping the wing, the drag at high speeds is reduced. This is covered in greater detail in Chapter 6. In flight near or above the speed of sound, swept wings are mandatory to reduce the power required to sustain cruise speeds. Most commercial transports, military aircraft, and newer business jets fly at near the speed of sound. Thus they require swept wings. A glance at the wing of a jet tells you how fast it is designed to go.

There are other reasons to sweep a wing. Swept wings impact stability. A wing swept back is generally more stable than a wing without sweep. This is desirable for passenger airplanes, since the airplane will have a tendency to stabilize after it is upset by a gust of air. Conversely, a wing swept forward will be less stable. Experimentation with forward-swept wings has resulted in less stable airplanes, which increases maneuverability. Although not yet found in production airplanes, the fighter of the future may employ forward-swept wings.

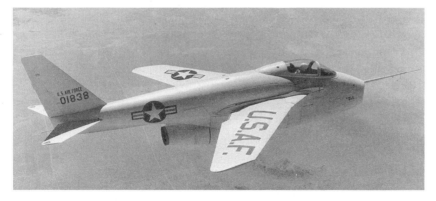

Fig. 3.7. The variable-sweep X-5 research aircraft. (*Photo courtesy of NASA.*)

Another reason to sweep a wing might be to move the center of lift of the wing forward or aft from where the wing root attaches to the fuselage. This might be necessary to accommodate a certain structural feature at the wing root and a certain position for the center of gravity of the plane.

DC-6 publicity photos used a model named Norma Jean Baker inside the cabin. She later became known as Marilyn Monroe.

Virtually all aircraft that have swept wings have wings swept back, though it is interesting to note that the high-speed drag reduction due to sweep is equal for both forward- and backward-swept wings. The primary advantage to forward sweep is to increase maneuverability. But forward sweep is very difficult to build structurally. A problem known as *structural divergence* can occur if the wing is not stiff enough. What happens is that the forward tip twists due to the load. The twisting increases the tip loading, thus twisting the wingtip even more. Eventually, the tip load is so great that it literally twists off the airplane. The advent of modern composites that can be tailored to specific stiffness requirements has led to the possibility of forward-swept wings. Composites can be manufactured in such a way that it eliminates the structural divergence. These airplanes are still very much in the experimental stage, however.

In the late 1970s NASA started a program to study swept-forward wings. The program resulted in the X-29, forward-swept wing demonstrator seen in Figure 3.8. The purpose behind the forward-swept wing was to create a naturally unstable platform. This is highly desirable for maneuverability. The airplane is so unstable that the pilot alone cannot control it. A sophisticated computer control system, called fly-by-wire, is added into the control loop to make the airplane controllable. If the computer were to fail, the airplane would instantly become unstable and crash.

Taper

The designer may consider a tapered wing for an airplane. A tapered wing has a shorter chord at the tip than at the root, as shown on a

Fig. 3.8. The X-29 with its forward-swept wing. (*Photo courtesy of NASA.*)

DC-8 in Figure 3.9. There are several advantages to tapering a wing. One reason to taper a wing is to adjust the load along the wing's span. The designer wants to distribute the wing loading such that it is reduced at the wingtip. A high tip loading will put a large bending load on the wing. This means that the entire wing structure has to be built stronger, and thus heavier, to handle the load. If you hang a swing from the end of a tree branch, the branch will bend considerably. Thus, we know from experience to hang the swing closer to the trunk of the tree. A rectangular wing will have a large *tip loading* and so will require a stronger structure. A tapered wing reduces tip loading and thus results in a lighter wing structure.

The primary disadvantage of a tapered wing is that it is more difficult to build. Most small, inexpensive aircraft use at least part constant-chord wings. The wing of the Cessna 172, shown in Figure 3.10, is a good example of a part constant-chord wing. A well-known example of a constant-chord wing is the 1960s line of Piper airplanes called the Cherokee, shown in Figure 3.11. These early Cherokees had a wing that was dubbed the "Hershey Bar" wing since it looked like a Hershey chocolate bar.

On June 13, 1979, a human-powered airplane, the Gossamer Albatross, crossed the English Channel. On July 7, 1981, the solar-powered Solar Challenger crossed the English Channel.

Fig. 3.9. The DC-8 illustrating wing taper. (*Photo courtesy of NASA.*)

Fig. 3.10. Cessna 172. (*Photo courtesy of the Cessna Aircraft Co.*)

There is another reason to taper a wing, and that is to adjust the lift along the wing to minimize the drag. As discussed in Chapter 2, the lift on a wing is proportional to the amount of air diverted by the wing times the vertical velocity of that air. The amount of air

Fig. 3.11. Piper Cherokee 140 with a "Hershey Bar" wing. (*Photo courtesy of Albert Dyer.*)

Fig. 3.12. The Republic P-47 with an elliptic wing shape. (*Photo courtesy of the U.S. Air Force Museum.*)

diverted, and thus the lift, changes along the span. In classical aerodynamics it can be shown that the lift of the most efficient wing tapers from the root to the tip in an elliptical fashion. It can be shown that keeping the downwash velocity, and therefore the angle of attack of the wing, constant along the trailing edge minimizes the induced drag. So the elliptical-lift distribution is obtained by shaping the wing so that the amount of air diverted tapers in an elliptical fashion. This theory became widely known to designers in the 1930s and resulted in many aircraft with elliptical wings. Two notable WWII-era fighters with an elliptic wing planform are the British Supermarine Spitfire and the American P-47 Thunderbolt shown in Figure 3.12. Today one does not see elliptic wings on airplanes because they are expensive to build and there are other ways to create approximate elliptical lift distributions.

Figure 3.13 shows the wing loading for a rectangular wing, an elliptic wing, and a linearly tapered wing. The rectangular wing has the highest loading at the tip, while the linear taper "unloads" the tip. A linearly tapered wing loses only 7 percent of the lift of an elliptic wing. Many airplanes use linear-tapered wings because of the reduced construction costs with only a small performance penalty.

Twist

One method of tailoring the lift distribution on the wing is to twist the wing, with the angle of attack greater at the root than at the tip. Another term for this type of twist is *washout*. The lower angle of attack at the tip unloads the tip and thus approaches the elliptical lift distribution. There is another advantage of using washout. Because the wing root is at a higher angle of attack, the wing will stall at the root first. Since the ailerons, which control roll, are usually on the outboard portion of the wing, the ailerons can still be effective after the

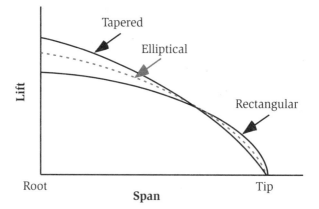

Fig. 3.13. The effect of wing shape on the distribution of lift.

root stalls. If the tip stalled first, the pilot would lose roll control during the stall. This could lead to an uncontrolled stall-spin. If wings are designed such that the root stalls first, the pilot can control the airplane and prevent the spin.

Though *mechanically* twisting the wing is common on general-aviation aircraft, commercial transports and other high-performance airplanes use an *aerodynamic twist*. Aerodynamic twist results when the wing sections are changed from root to tip. In other words, different wing section designs are selected for different positions along the span of the wing. For example, a designer might reduce the camber of the wing sections from root to tip. The goal is to select a wing that is more lightly loaded at the tip and will not stall at the same angle that the wing root stalls. The result is that the wing behaves as though it is twisted, although it is twisted aerodynamically and not mechanically.

Philanthropist Daniel Guggenheim was an aviation enthusiast. The Daniel Guggenheim Fund established seven aeronautical schools in the late 1920s. These were MIT, Caltech, University of Michigan, NYU, Stanford, Georgia School of Technology (now Georgia Tech), and the University of Washington. Of the seven original, only NYU no longer has a program in aeronautics.

Wing Configuration

There is still more to a wing than its airfoil and planform. The configuration of the wing, as viewed from the front, can affect stability, efficiency, and practicality. Why are wings sometimes slanted up or down from the root to the wingtip? Should the airplane have a low wing, a high wing, or a midwing? What type of wingtip is best? When should one consider a biplane?

Dihedral

Roll and yaw stability are desirable characteristics for trainers and transports, both of which can be enhanced by adding *dihedral* to the wings. Dihedral is the upward angle of the wing along the span against the horizon, as shown in Figure 3.14. Many are taught that the reason dihedral adds stability is that as the airplane rolls gravity pulls the upward wing back to horizontal. Unfortunately, the truth is more complex.

When an airplane enters a roll, there is a tendency to yaw in the opposite direction. For example, if the airplane rolls to the right, it is accompanied by a yaw to the left. That is, the

Manfred Von Richthofen (the Red Baron) flew the Fokker triplane, for which he is famous, for only 6 weeks and 19 of his 80 victories.

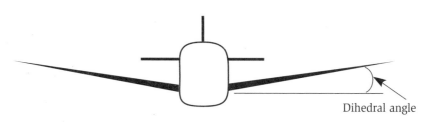

Dihedral angle

Fig. 3.14. The dihedral angle.

nose of the airplane swings to the left. This is called *adverse yaw*. This is why the pilot must use compensating rudder to make a *coordinated turn*. The reason for adverse yaw is that the wing rotating upward experiences more drag than the wing rotating down. Whether the rotation was caused by the ailerons or by a gust of air, the airplane rotates because one wing has more lift than the other. We know that lift requires work and this reflects itself in increased induced drag. Thus the wing with the greatest lift has the greatest drag.

> The Wright brothers No. 3 glider of 1902 initially had no rudder. During longer gliding flights they practiced turns. Much to their dismay when they banked the airplane the result was that the airplane went into a "skid" because the nose turned opposite to the bank. They had discovered adverse yaw. The quick-minded brothers deduced the need for a vertical stabilizer and rudder and thus completed the three-axis control puzzle that made controlled flight possible.

Dihedral adds stability because the adverse yaw results in a net reduced angle of attack on the upper wing, and an increased angle of attack on the lower wing. This results in an increase in lift of the lower wing and a reduction in lift of the upper wing, as well as a restoring yaw force. The result is a tendency to return to straight-and-level flight. Small general-aviation aircraft and commercial transports all have dihedral. These airplanes tend to return to level flight after gusts or accidental control inputs. Today airplanes are so stable cross-country trips can be quite boring.

During the early days of flight the Wright brothers advocated a design in which the airplane was slightly unstable. They felt that this forced the pilot to pay attention and continually fly the airplane. Meanwhile, competitors built airplanes that were inherently stable. Eventually, the Wright brothers could not compete with these stable designs. This is one factor that has led to the eventual loss of the Wright name in any airplane company today.

For completeness we should mention another stabilizing effect of all wings. Consider an airplane that is rotating because of a gust of air or a sudden short movement of the ailerons. After the gust of air is over or the ailerons are back to neutral, the lift on both wings would seem to be the same, though the inertia keeps the airplane rotating. But, while the airplane is in rotational motion, the wing going up sees the relative wind with a reduced angle of attack, and the lower wing sees an increased angle of attack. This causes a force opposite to the rotation and tends to bring the rotation to a stop. When the rotation stops, the counterrotational force also ceases.

Some military aircraft, such as the F-104 (Figure 3.15), have *anhedral* wings. Anhedral is negative dihedral; in other words the wings slope down as seen in the picture. Just as dihedral is stabilizing, anhedral is destabilizing. As mentioned earlier, stability reduces maneuverability, so by making the airplane less stable it becomes more maneuverable. Anhedral is found only on military fighters and aerobatic airplanes.

High Wings vs. Low Wings

The wing designer must also decide where to put the wing on the airplane. Should it be a low wing, high wing, or midwing? What are the benefits of wing position? The effects of wing position on stability are fairly minor. A high wing has a little more stability than a low wing, but the change in stability is small when compared to the effects of sweeping the wing and adding dihedral. So, the choice of wing position is based on much more practical matters. Private pilots frequently argue this point in hangar talks.

Howard Hughs's infamous Spruce Goose was actually made mostly of birch.

Fig. 3.15. The F-104 with anhedral (negative dihedral). (*Photo courtesy of NASA.*)

Those who fly low-wing airplanes insist that the clear view of the sky outweighs the lack of visibility downward because of the wing. The high-wing pilots argue that the view of the ground is more important than the view of the sky. Those who fly in hot, sunny regions also appreciate the high wing for shading the cabin.

Besides a slight increase in stability, a high-wing aircraft offers the ability to locate the fuselage close to the ground. Military transports use this configuration so that equipment can be easily loaded and off-loaded. A good example is the C-130 Hercules shown in Figure 3.16. High wings also offer more room for high-lift devices. For example, wing flaps can extend farther down without a concern for ground interference. Another advantage is that wing struts can be used under the wing where they will not interfere with lift.

A WWII DC-3 lost a wing from a bomb while on the ground. The only available replacement was a DC-2 wing, which was 5 feet shorter and designed for a much smaller load. The wing was attached and the airplane, dubbed the DC-2.5, flew to safety.

A disadvantage of high wings is that the landing gear must be placed in the fuselage. This usually adds bulging pods to accommodate the gear. Another disadvantage is that the low fuselage leaves little tail clearance. In order for the airplane to be able to rotate at takeoff the fuselage must have upsweep, as shown in the picture. This sacrifices valuable cargo space.

Low-wing airplanes make landing gear placement much simpler. However, for multiengine airplanes with engines hung on the wings, the landing gear must be long enough to prevent ground interference. Low-wing airplanes also have structural benefits, since the

Fig. 3.16. The C-130 Hercules demonstrating fuselage upsweep. (*Photo courtesy of Paul Burke.*)

wing *spars* (the internal beam supporting the wing) can carry through the lower fuselage below the passenger deck, resulting in a continuous spar structure. This allows the wing to be fully *cantilevered* (supported by one end) with no need for external bracing. Most commercial transports since the DC-3 have chosen this approach. Because the fuselage is positioned higher than for a high-wing airplane, there is less need for fuselage upsweep. The primary disadvantage is the requirement for long landing gear to provide ground clearance for engines.

Many did not believe the Wright brothers had flown in 1903. Until their demonstration in France in 1908, the French called them the "Wright liars."

COWLING SAVES THE BOEING 737

The Boeing 737 was initially designed with turbojet engines (see Chapter 5). To improve efficiency the move was made to turbofans engines, which have a bigger diameter. The Boeing 737 had been designed close to the ground since the small-diameter turbojets did not need much clearance. Boeing had difficulties retrofitting the Boeing 737 with turbofans engines with so little clearance. The result is an unusual cowling design (see Figure 3.17) that is flat near the bottom. The insert in the picture is of the old engine for comparison. Creating this innovative cowling saved the Boeing 737, which has now become the most popular commercial jet airplane in history.

Fig. 3.17. The Boeing 737 noncircular engine inlets with the old engine as insert.

Airplanes with wings positioned in the middle of the fuselage are usually found only in military fighter or aerobatic airplanes. The midposition offers the benefit of clearance for stores underneath the wing, while maintaining visibility above. The configuration also has the lowest drag. The primary disadvantage of a midwing configuration is that the wing-fuselage joint occurs in the middle of the fuselage. The structure must carry through the fuselage at a point where one might want to put passengers, cargo, etc.

Wingtip Designs

The design of the wingtip has an effect on lift. A nice rounded wingtip looks clean but actually results in lower performance than a square tip. The design of the wingtip also affects the distribution of lift along a wing and thus the wing's efficiency. Thus the shape of the wingtip has become an important design criterion.

Choosing the correct wingtip is a matter of compromises, which include aerodynamic performance, structural loads, manufacturing, and (possibly most important) marketing. In the early days rounded tips were simple to build because all the builder had to do was bend a rod from the leading edge around to the trailing edge. Since loads are small at the wingtip, little internal structure was needed. Many of the pre-WWII airplanes have rounded wingtips. But, squared-off tips are aerodynamically more efficient. The squared-off wingtip better restricts the passage of high-pressure air from the lower surface to the upper surface. Any high-pressure air leaking to the upper surface leads to lower aerodynamic efficiency. Thus, most of today's airplanes have simple squared-off tips.

> The Boeing 777 was designed with folding wingtips. No customer ever ordered them.

Winglets

Today, many aircraft sport *winglets,* which are wingtips turned vertically, as shown in Figure 3.18. While the winglets point up on wings, it should be noted that winglets on horizontal stabilizers point down since the horizontal stabilizer pulls down. Winglets go one step further in preventing the passage of high-pressure air from flowing around from the lower surface to the upper surface. In essence, they provide a block. The result is that the wing can carry a finite amount of lift all the way to the tip. As we discussed in Chapter 2, the efficiency of a wing increases with length. Winglets increase the effective length of the wing and thus increase the wing's efficiency without increasing its length. Winglets have become the golden child of all airplane sales representatives.

However, winglets do come with disadvantages. The net effect of the winglet is roughly equivalent to laying the winglet flat, which would result in extra span and wing area. So, one could achieve the same wing performance by doing exactly that, laying the winglet flat. The added tip loading requires a stronger and heavier wing. So, winglets cannot be retrofitted to existing airplanes without changing operating conditions to lower the wing loading. But, on a new design, winglets can help make the wing more efficient by reducing the induced power required for a given lift, just as extra span would. Nevertheless, one of the biggest reasons for the preponderance of winglets on today's business jets is that they are considered very sexy.

Northrop invested $1.2 billion in the early 1980s to develop the F-20 Tigershark. Despite outstanding performance, not a single one was sold.

One might wonder why the winglets of some of the jets cover only the last part of the wingtip, as in Figure 3.18. This is because, at cruise speeds, the air flowing over the first part of the wing is going at or near the speed of sound. A winglet here would cause a shock wave and increased drag, which is discussed in Chapter 6.

Canards

The Wright brothers designed their airplanes such that the horizontal stabilizer was forward of the wing. This configuration is called a *canard,* which is the French word for a type of duck. But the canards soon disappeared from the scene, until their recent revival. Today, canards have become popular, much like wingtips, because of their sexy appearance. A dramatic example of a canard is the

Fig. 3.18. A winglet on a K-C135A. (*Photo courtesy of NASA.*)

NASA X-29 experimental aircraft shown in Figure 3.8 and the Long-EZ in Figure 3.19.

Canards have the advantage that the horizontal stabilizer is lifting up rather than down, as it does on a conventional airplane. This reduces the load on the wing. Thus, the canard appears to be more efficient. Both the wings and the horizontal stabilizers provide lift. But canards must be designed such that the horizontal stabilizer stalls before the wing. If the wing stalled first in the canard configuration, the rear of the airplane would drop, increasing the angle of attack further, and stall recovery would be impossible. The canard is designed such that the horizontal stabilizer stalls first, dropping the nose of the airplane. Thus, the airplane's wing will not stall. This safety feature has encouraged many modern designers to favor canards.

Canards are touted as being more efficient than conventional airplanes because all surfaces are lifting. This conclusion ignores power. The horizontal stabilizer is at a higher angle of attack so it is working harder than the wing. By comparison, the wing is loafing. The lift from the horizontal stabilizer is generated by increased downwash because of the higher angle of attack. From a power standpoint, this is inefficient. The more efficient wing is allowed to generate less lift, as a result of the load carried by the canard. For

Fig. 3.19. A home-built Long-EZ illustrating a canard configuration. (*Photo courtesy of Sandy DiFazio.*)

efficiency it would be better to have the wing working harder. So, the canard sacrifices efficiency so that the horizontal stabilizer always stalls first.

Choosing a canard configuration over a conventional configuration has to be considered carefully. The total power required for lift must be analyzed before the designer decides whether canard or conventional configurations are best.

Boundary-Layer Energy

The concept of the boundary layer was introduced in Chapter 2. In order to understand certain aspects of wing design, it is necessary to go into more detail. In this section the effects of pressure and energy on the boundary layer will be discussed as an introduction to high-lift devices.

The boundary layer on a wing section is shown in Figure 3.20. In this figure the boundary-layer thickness is greatly exaggerated. In reality, the boundary layer is quite thin. For example, at the trailing edge of a Boeing 747-400 (where the boundary layer is the thickest) the boundary layer is approximately 1 inch (2.5 cm) thick. Figure 3.21 is a blowup of the boundary layer showing how the speed relative to the wing changes

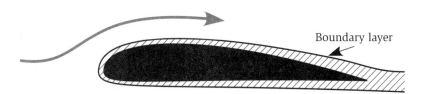

Fig. 3.20. The boundary layer.

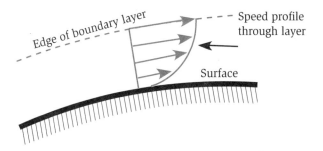

Fig. 3.21. How the airspeed changes in the boundary layer.

from zero at the wing surface outward. It is the friction of air molecules with the surface that causes this change in speed.

The lower speed near the surface translates into lower kinetic energy. Boundary-layer energy is important because higher kinetic energy will allow the boundary layer to continue to follow a surface even when the surface curves away. This is the essence of the Coanda effect mentioned earlier. As stated in Chapter 2, most of the lift on a wing is in the first fourth of the chord length. This is where the pressure is lowest. As the air moves back along the top of the wing, the pressure increases until it reaches the ambient pressure at the trailing edge of the wing. This is known as the *trailing-edge condition*. Thus, the air is moving into an increasing pressure, which tends to slow it down. If the boundary layer has enough energy to overcome the increasing pressure, it will follow the wing's surface. When the energy of the boundary layer is not sufficient, the boundary layer will stop flowing and separate from the surface. Past the separation point the wing experiences air flowing in the reverse direction. The wing is entering a stall. Separation usually occurs near the trailing edge at the critical angle of attack. As the angle of attack increases, the point of separation moves forward and lift decreases.

In March 1945 a C-47 (military version of the DC-3) had its left wing severed off just outboard of the left engine from a midair collision. The pilot managed to make a controlled crash landing with only one wing.

An understanding of the energy in the boundary-layer air is necessary for wing designers to design wings that hinder separation. If a wing can reach a higher angle of attack before stalling, it will be able to take off and land at lower speeds or carry heavier loads. Lower stall speeds translate to shorter runways, and heavier loads translate to greater revenue.

ICE ON A WING

A wing designed to stall from the trailing edge first may lose this characteristic when it flies into icing conditions. Ice forms on airplanes that fly into moisture in a certain temperature range. Supercooled water drops freeze on impact and form rime ice, which is rough and opaque. Water that is warmer will impact the wing and form glaze ice, which is smooth and clear. Mixed ice is a mixture of glaze and rime ice.

ICE ON A WING *(Continued)*

The buildup of ice does several negative things to the wing. The first is that it changes the shape of the wing and thus changes its stall characteristics. In general, a wing with ice will stall abruptly at a lower angle of attack than without ice. Ice also adds weight, which is an additional load on the wing. Therefore, to compensate, the pilot must increase the angle of attack. Eventually, the angle of attack required to maintain flight reaches the new stall angle of attack and the airplane can no longer fly. An additional impact of ice is increased drag, both induced and parasitic. The result is that the power requirement increases.

Boundary-Layer Turbulence

Normally, as passengers or pilots, we associate turbulence with atmospheric conditions. This is known as *clear-air turbulence.* Clear-air turbulence is important to structural designers to design for gust loading. But there is another type of turbulence that is critical to the aerodynamic design of the wing. This is *boundary-layer turbulence,* which occurs only in the very thin boundary layer.

For a week every summer the busiest airport in the world is Whitman Field in Oshkosh, WI. This occurs during the Experimental Aircraft Association (EAA) annual fly-in.

Most pictures of wings with air flowing over them show smooth patterns of air. This is *laminar* flow, which is nonturbulent flow. Figure 3.22 illustrates both laminar and turbulent flows. In the picture smoke in laminar-flow air passes through a special screen. Shortly after the screen the laminar flow becomes turbulent. The basic difference is that the laminar flow is smooth while the turbulent flow is chaotic. Skin friction, and thus skin drag, increases dramatically with boundary-layer turbulence. A wing with laminar flow will have much less skin drag than a wing with turbulent flow.

It would be a great asset to design a wing that was laminar over its entire surface. But this has proved extremely difficult. Boundary-layer turbulence is a natural phenomenon and it becomes more prevalent as speed and the size of the airplane become greater. Besides this natural

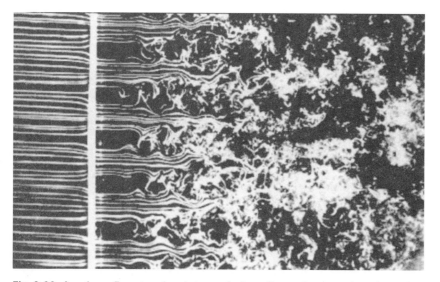

Fig. 3.22. Laminar flow turning into turbulent flow when passing through a plate with holes. (*Photo courtesy of Thomas Corke and Hassan Nagib.*)

tendency of a laminar boundary layer to become turbulent, things such as rivets, bugs, and raindrops can trigger boundary-layer turbulence. A great deal of effort has gone into understanding boundary-layer turbulence and devising ways to reduce or eliminate it. Modern wings attempt to maintain laminar flow as far back on the wing as possible but transition to turbulence before the air has a chance to separate. If you look at the wing of a jet, you will see that it is smooth part of the way back and then there is less concern for smoothness.

Turbulence does have one beneficial effect. Because the airflow in the boundary layer is churning, it mixes the most energetic air with air close to the wing surface. Thus the boundary layer is *energized* and has increased kinetic energy. The advantage of this is that the air will stay attached to the surface longer. A laminar-flow wing will stall at a lower angle of attack than a wing with turbulent flow. Another method for *energizing* the boundary layer is to add *vortex generators*. Many airplanes use vortex generators to delay separation of the air, and thus delay stall. Vortex generators are discussed below.

A popular home-built airplane has had problems flying in rain. The Vari-EZ and Long-EZ, shown in Figure 3.19, have laminar-flow canards. When flown in rain, the water droplets ruined the laminar-flow properties. The aerodynamic changes seen by the canard were sufficient for pilots to notice changes in the handling characteristics.

Form Drag

We should look at drag a little more carefully in order to better understand stalls. Parasite drag is composed of two parts. In Chapter 2 the effect of friction was discussed. Also mentioned was form drag. Form drag is the drag associated with moving things like antennae and wheels through the air. Form drag can be thought of as the drag associated with pulling the wake of the airplane along with it. Form drag plus skin drag make up parasite drag.

Form drag is what people usually associate with aerodynamic shapes. People often look at cars and instinctively know which are better aerodynamically. We tend to look at how streamlined the car is. What usually distinguishes a streamlined car from an unstreamlined car is the apparent wake resulting from the rear of the car. A truck clearly has a large wake and high form drag, as illustrated in Figure 3.23. It is desirable to reduce form drag on an airplane as much as possible. This means reducing the wake. Any part of an airplane where the air separates from the surface produces a wake. Even a small, cylindrical antenna will produce a significant wake. Therefore, protuberances such as antennae are encased in aerodynamic fairings.

The wing of a Boeing 747 has the same parasitic drag as a 1/2-inch cable of the same length.

Until about 1920, airplane designers thought wings had to be thin. To create an efficient structure, there was much use of external wire bracing. Ironically, the external bracing resulted in much higher drag than what was saved by making the wings thin. Late in WWI the Germans discovered that the wire bracing was adding too much drag and started to use fatter wings that could hold more structure.

Fig. 3.23. Form drag illustrated by the wake of a truck.

In discussing the stall of a wing, we have focused on the reduction in lift. Almost as great a problem is the wake caused by a stalling wing producing significant form drag. Just when the airplane needs all the power it can get to recover from the stall, there is a large additional power requirement put on the airplane, making the recovery even harder.

THE GOLF BALL

Turbulence can have a beneficial effect on reducing the wake of an object. If the surface air in the boundary layer becomes turbulent, the higher energy in the turbulent flow will help the air stick to the surface longer before separating. The result is lower form drag.

A smooth ball traveling through the air will have high form drag, as shown in Figure 3.24. The laminar flow soon separates, producing a wake behind the ball that far outweighs the skin drag. One solution to improve the range of a ball is to energize the air around the ball by churning up the air. This will allow the air to remain attached to the surface longer and will reduce the size of the wake.

A golf ball's size would normally result in laminar flow and a sizable wake. That is why the surface is covered with dimples to encourage turbulence. As seen in the figure, the dimples reduce the size of the wake and thus reduce the form drag. The result is that golf balls with dimples travel much farther than if they were smooth.

Vortex Generators

A boundary layer with more energy will be less prone to separate. *Vortex generators* are a means for adding energy to the boundary layer and thus increasing the stall angle of attack. Vortex generators are simple devices that can be retrofitted to any airplane. They are simply small wings set at an angle of attack to the local air, as shown in Figure 3.25.

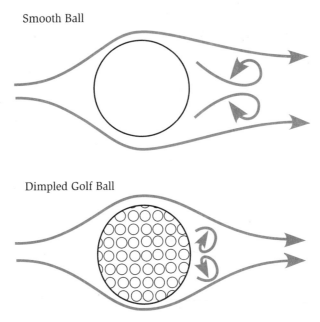

Smooth Ball

Dimpled Golf Ball

Fig. 3.24. The dimples on a golf ball decrease form drag.

Since a wing at an angle of attack generates a trailing vortex sheet, so do vortex generators. So, for vortex generators to work they only need to be as high as the boundary layer is thick. They are usually added a little aft of the region of highest air velocity.

Unfortunately, there is a penalty for adding vortex generators. Since vortex generators put a force on the air, they do work, yet without adding to the lift of the wing. Therefore, additional power is needed when vortex generators are used. The reality of vortex generators is that they are used most often to correct an existing problem. Because they introduce a power penalty, there is no incentive for putting them in the original wing design. A well-designed wing should not need vortex generators.

In 1911, Orville Wright made a 9-minute flight in a glider. It remained a world record for 10 years.

Nevertheless, there are several cases where it is desirable to utilize vortex generators. The first is in situations where the vortex generators are used only for takeoff and landing but are stowed for cruise. A particular situation where this can be seen is on the leading edge of some wing flaps. The use of flaps is discussed in the next section.

Fig. 3.25. Vortex generators on the slat and wing of a Navy A-4 Sky Hawk. (*Air Classics Museum, Auora. IL.*)

The second case where vortex generators are desirable is when an existing airplane is modified to meet new performance parameters. A typical change in performance requirements is to make an existing airplane use less runway for takeoff and landing. Ideally, a new wing would be designed. However, designing a new wing for an existing airplane is usually too costly. Vortex generators are thus added to increase takeoff and landing performance. There are many companies that modify existing airplanes, converting them to STOL (short takeoff and landing) airplanes, which is the case in Figure 3.25.

High-Lift Devices

Devices or modifications to the wing that increase the stall angle of attack are called *high-lift devices*. Airplanes employ various high-lift devices to improve takeoff and landing performance. The basic principle behind these devices is that they allow the wing to divert more air down without stalling. Vortex generators discussed above are a type of high-lift device. During takeoff and landing the airplane needs to fly at the lowest speed possible, since high takeoff and landing speeds mean longer runways. Therefore, a major goal in designing a wing is to reduce the stall speed as much as possible. The easiest way to do this would be to design the wing with a great deal of wing area and camber. Such a wing would be able to fly at a high angle of attack without stalling. But as mentioned before, a high-cambered wing would have a high drag at cruise speeds. The solution is to design a wing that can change its characteristics for takeoff and landing speeds. A second way to increase the stall angle of attack, discussed above, is to energize the boundary layer.

The most common high-lift device is the wing flap. The next most common is to add leading-edge devices called slots and slats. In rare instances the deflected slipstream from the propellers or jet engine is diverted to provide additional lift at low speeds.

In a rush to be the first to fly nonstop from New York to Paris, Charles Lindbergh set a record with a nonstop overnight flight from San Diego to St. Louis.

Flaps

Wing flaps can be found on virtually every modern airplane. The effect of adding flaps to the trailing edge of the wing is equivalent to increasing the camber of the wing. Some flap designs also increase the chord length of the wing. This increases the area of the wing so that more air is diverted, thus reducing the angle of attack needed for lift.

There are many types of flaps. In the 1930s and 1940s the *split flap,* shown in Figure 3.26 was introduced and was one of the first types of flap to appear in production airplanes. Splitting the last 20 percent or so of the wing forms this type of flap. The top surface of the wing does not move while the bottom surface lowers. The split flap is effective in improving the lift, but it creates a great deal of form drag, as shown in the figure. The split flap was used on the DC-3. It was also used on WWII-era dive-bombers because it helped increase lift at low speeds and slowed the airplane during the dive.

The simple *hinged flap* (Figure 3.27) is most common on smaller aircraft. The last 20 percent or so of the inboard section of the wing is simply hinged so that it can increase the camber. The first 20 degrees of flap extension increase the lift without greatly increasing the drag of the wing at low speeds. Many airplanes extend their flaps to 10 or 20 degrees on takeoff in order to shorten the takeoff distance. When the flaps are extended greater than 20 degrees, the form drag increases rapidly with little or no increase in lift. Increasing the drag increases the descent rate, which is desirable during the approach

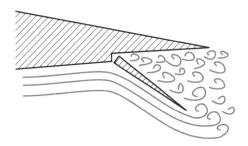

Fig. 3.26. Split flap.

for landing. Thus it is not uncommon for an airplane to land with the flaps set at 40 degrees.

A more sophisticated flap is the *Fowler flap* shown in Figure 3.28. With the Fowler flap, the rear section of the wing not only changes angle but also moves aft. The result is both an increase in camber and an increase in wing area. A bigger wing will divert more air and increased camber will increase the downwash velocity. Mechanisms to operate Fowler flaps can be quite complicated.

The maximum lift that a flap can generate is limited by the critical angle where the flap begins to stall. This has been improved by the introduction of *slotted flaps.* A single slotted flap is shown in Figure 3.29. A slotted flap extends both aft and downward, like a Fowler flap,

In 1921, Bessie Coleman was the first African-American woman to receive a pilot certificate. She had to go to France to obtain it, since she was not allowed to in the United States.

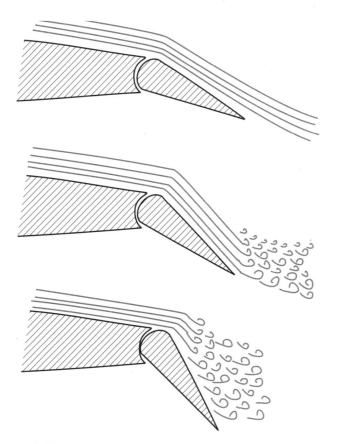

Fig. 3.27. Hinged flap.

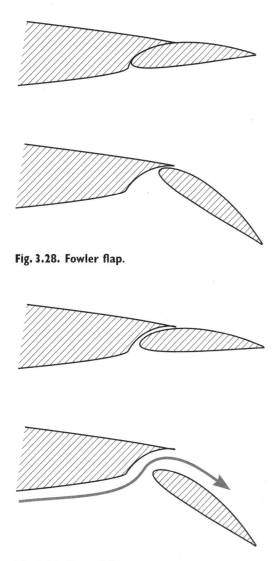

Fig. 3.28. Fowler flap.

Fig. 3.29. Slotted flap.

In 1991, the very first Boeing 727, the Spirit of Seattle, was retired after 64,492 flight hours, which is equivalent to 7.4 years in the air.

plus it is designed to take advantage of the gap between the flap and the wing. The air in the boundary layer, having passed over the top of the wing, has lost most of its kinetic energy. Thus when it reaches the extended flap it is likely to separate from the flap and cause a stall. However, the air passing under the wing does not face the same problem. The slot between the wing and the flap diverts some of the higher-energy, lower-surface air to the top of the flap. The air remains attached to the flap longer, thus reducing drag and inhibiting stalls. A *double-slotted flap* (Figure 3.30) basically repeats this step twice, using two separate flaps in tandem. This provides the maximum lift from a flap design. The disadvantage of this design is that the operating mechanism is very complicated and heavy. Multislotted flaps are seen on many modern passenger jets, while large airplanes use single-slotted flaps.

Until the 1990s airplane performance was the key design criterion. Airplane companies were proud of sophisticated triple-slotted flap systems. During the 1990s a shift toward reducing cost as a key design criterion has pushed airplane companies to maximize the performance of single-slotted flaps. One technique that is used is to place vortex generators on the leading edge of the single-slotted flap. When the flap is retracted, the vortex generators on the flap are hidden in the wing. Thus, the vortex generators do not penalize the airplane in cruise but are available for takeoff and landing.

The next time you fly a commercial airplane ask for a window seat behind the wing. During the approach and landing phase of the flight, watch the wing unfold. It is truly remarkable how the wing evolves into a high-lift wing from its normal cruise configuration.

> With the design of the Boeing 747, many felt that the ultimate wing had been built. The wing boasted triple-slotted flaps and was the pride of the engineers who designed it. Indeed it is a marvel of performance. Unfortunately, building a triple-slotted flap is costly. In the new paradigm where cost is one of the leading design constraints, the goal is to design the simplest (in other words cheapest) flap possible. Airbus has a single-slotted flap on their Airbus A320.

Slots and Slats

Leading-edge devices, like flaps, are sometimes used to increase the camber of the wing and increase the stall angle of attack. But the details are somewhat different. Other times, the purpose of the leading-edge devices is much like that of the slot in a slotted flap. These devices allow the high-energy air from below the wing to flow to the upper surface of the wing. This energizes the boundary layer. Thus, the wing stalls at a higher angle of attack and the maximum lift is increased.

The simplest leading edge device is the *fixed slot* shown in Figure 3.31. This is a permanent slot near the leading edge of the wing. The high-pressure air below the wing is drawn up through the slot and flows over the top of the wing. This energizes the boundary on top of the wing. A permanent slot can increase the critical angle of attack significantly. The disadvantage of the fixed slot is that it causes increased power consumption and drag at cruise speeds.

A device similar to the slot is the *fixed slat,* shown in Figure 3.32. It is added onto the wing, increasing the wing's cord length as well as energizing the boundary layer. Like the fixed slot, the fixed slat causes increased drag at cruise speeds.

The solution to the drag caused by fixed slots and slats is to design a slat that is deployed only at slow speeds and causes little or no drag in cruise. The *Handley-Page retractable slat,* shown in Figure 3.33, extends to large *droop angles* to give the wing large leading-edge

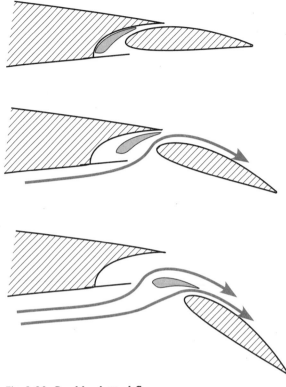

Fig. 3.30. Double-slotted flap.

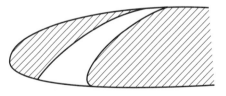

Fig. 3.31. Fixed slot.

camber. In cruise the slats are retracted and do not cause increased drag. This type of slat is often designed so that they deploy by themselves at slow speeds and high angles of attack and return to the flush position in cruise.

Deflected Slipstream and Jet Wash

One way to increase lift at slow flight speeds is to divert the propeller's slipstream or the jet engine's exhaust down. To achieve a substantial

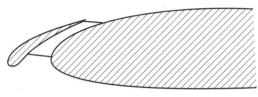

Fig. 3.32. Fixed slat.

Fig. 3.33. Handley-Page retractable slat.

lift increase with a slipstream, the plane must have engines mounted on the wings with large propellers that generate a slipstream over a substantial portion of the wing. The wing must also have a multislotted flap system to deflect the slipstream effectively. This technique has not found significant commercial applications.

The exhaust of a turbofan-powered airplane, which is described in Chapter 5, can be diverted down to produce additional lift at low speeds. One way to produce the diversion is to have the flaps extend down into the exhaust when fully extended. One problem with this technique is that the flap extension into the jet exhaust exposed it to very high temperatures, creating a significant design challenge.

Another way to divert the jet exhaust is to mount the engines on the top of the wing with the engine exhaust crossing the top of the wing as in Figure 3.34. Flaps behind the engines use the Coanda effect to divert the exhaust down when extended. This gives a substantial increase in lift for takeoff and landing.

In 1934, antitrust laws prevented airframe manufacturers from owning mail-carrying airlines. Boeing holdings under United Aircraft and Transport Corp. became three entities: United Air Lines, United Aircraft (later renamed United Technologies, of which Pratt & Whitney is part), and the Boeing Airplane Company.

Fig. 3.34. NASA QSRA STOL research vehicle with deflected jetwash. (*Photo Courtesy of NASA.*)

Wrapping It Up

Before moving to the next chapter discussing stability and control, let us look at some typical wings and identify what choices were made in designing them. We will look at the following three airplanes: the Cessna 172 from general aviation, the Boeing 777 from commercial aviation, and the Lockheed-Martin/Boeing F22 stealth fighter from the military.

The Cessna 172, shown in Figure 3.10, is a popular four-seat airplane with a cruise speed of 120 knots. The wing is unswept but has a small amount of taper toward the tip. The wing is mounted at the top of the fuselage for stability and for structural reasons. The wing has an aspect ratio of 7.5, dihedral for stability, hinged flaps for landing, and round leading edges for a gentle stall. The low wing loading gives it good low-speed performance, but its top speed is not particularly noteworthy. The airplane is designed to be easy to fly and sports the best safety record of all general-aviation aircraft.

The Boeing 777 (Figure 3.35) is designed to carry a heavy load over long distances, its high cruise speed, about Mach 0.84, requiring a high wing loading. The aspect ratio is similar to that of the Cessna 172, but the wing is swept and tapered. In this case the sweep is necessary for the high cruise speed. Taper, plus changes in wing section, result in a roughly elliptic lift distribution at cruise. For takeoff and landing, the airplane has double-slotted trailing-edge flaps and deployable leading-edge slats. The wing is fairly thick (a person can stand in the wing root) to accommodate the necessary structure and fuel. The wing also has significant dihedral for stability.

The Lockheed-Martin/Boeing F-22 is the current generation stealth fighter (Figure 3.36). The body and wing blend together for stealth. As a result it shares the characteristics of a low-wing and a midwing design. The wing is more highly swept and tapered than the Boeing 777. It is highly maneuverable and can cruise at speeds greater than the sound speed. The F-22 employs many high-lift devices, including diverting engine thrust down. Its wing loading is high and it has a very small aspect ratio.

Fig. 3.35. Boeing 777. (*Used with the permission of the Boeing Management Company.*)

Fig. 3.36. Lockheed-Martin YF-22 stealth fighter. (*Photo courtesy of the U.S. Air Force Museum.*)

In this chapter we considered the factors that go into the design of the wing. We also discussed the stability created by sweep and dihedral. This is called *lateral stability*. It is time to look at the stability in the other two axes and to better understand the meaning of stability. In the next chapter we do just that as we look at stability and control.

CHAPTER 4

Stability and Control

One of the greatest improvements in aircraft over the last few decades has been in the area of stability and control, which is discussed in this chapter. Stability is the tendency of an airplane to return to a previous condition if upset by a disturbance, such as a gust or turbulence. Control is the ability to command the airplane to perform a specific maneuver or to maintain or change its conditions.

Before WWII stability and control did not receive much emphasis. The issue was merely how well the pilot thought the airplane "handled." Different philosophies reigned. For example, the Wright brothers felt that a less stable airplane was better because they believed it forced the pilot to be diligent. The Wrights' competitor, Glenn Curtiss, believed that an airplane should be very stable to reduce the pilot's workload. The Curtiss camp prevailed in the long run.

After WWII, engineers started to develop quantitative means for determining airplane stability and handling properties. Today the computer can do the work of controlling the unstable airplane while the pilot focuses on other tasks. In these next few sections we guide you through some basic principles of stability and control.

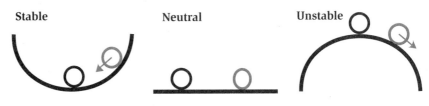

Fig. 4.1. Three types of static stability.

Static Stability

There are three types of static stability. These are statically *stable, unstable,* and *neutral.* All three are illustrated in Figure 4.1. The ball in the bowl illustrates a statically stable system. If it is displaced from the bottom, it will tend to return. An increase in the steepness of the sides of the bowl corresponds to an increase in stability. If one turns the bowl over, we have an illustration of static instability. If the ball is moved from the exact center of the bowl, it will continue to move away. The ball on the table illustrates neutral stability. If the ball is moved, it will tend to stay in its new position.

For an airplane, static stability means that if a gust of air or some other perturbation causes a change in its current state such as heading, it will experience a restoring force. Small general aviation and commercial aircraft, if properly trimmed, will return to straight-and-level flight after a gust or an abrupt disturbance of the controls. There have been instances where a pilot has fallen asleep only to have the airplane continue its course without an autopilot. Even Lindbergh got a short, unplanned nap on his solo trip across the Atlantic, and lived to tell about it.

Stability should not be confused with balance. An airplane is balanced when there is no net torque on the airplane. Stability is the tendency for the airplane to return to its previous attitude once disturbed. In the next section we lead you through pitch stability, which should make this distinction clear.

Longitudinal Stability

Longitudinal stability is the tendency for an airplane at a specific pitch attitude to return to that attitude when perturbed. Ultimately, we are interested in the stability of the airplane as a whole. However, to introduce the concept of stability we first focus on the longitudinal stability of a wing. Only after the stability of a wing is understood will you be introduced to the purpose of the horizontal stabilizer and stability of the airplane as a whole.

Stability of a Symmetric Wing

We have chosen to use a symmetric wing section for this example to simplify the explanation. The description does not change for a nonsymmetric wing, but a few additional ideas would have to be introduced, which would only confuse the issue.

The wing has a center of gravity (c.g.) as marked in Figure 4.2. The center of gravity is where the wing balances. If you could combine the distributed weight of the wing into a single point, the location of this point would be the center of gravity, which is also at the center of balance. Similarly, the center of lift is the point where we would place the lift if we took the distribution of lift over the wing and placed it at a single point.

Figure 4.2 shows three situations, a wing that is stable, neutrally stable, and unstable. For the stable wing the center of gravity is ahead of the center of lift. Let us say that a gust of air increases the angle of attack. The greater angle of attack will increase the lift. This increase in lift will cause the wing to rotate about the center of gravity and reduce the angle of attack. In other words, there is a rotational torque that rotates the wing back in the direction from where it started. So, we say this wing is stable.

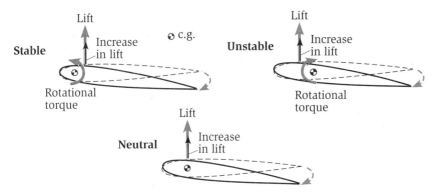

Fig. 4.2. Stability of a symmetric wing.

For the neutrally stable wing the center of gravity and the center of lift are colocated. So, if the wing is disturbed by a gust, increasing the angle of attack, the lift will increase but will not result in any restoring torque around the center of mass. Since there is no rotational torque, this wing will tend to remain just where the gust puts it. Thus it is neutrally stable.

The unstable wing has its center of gravity behind the center of lift. Now, if a gust increases the angle of attack, the lift increases and there is a net rotational torque, which works to increase the angle of attack even further. Any disturbance will be amplified. This situation is unstable.

Balance

The above description illustrates stability. As mentioned before, stability is the tendency to rotate back in the direction toward its initial state. Balance, on the other hand, occurs when there is no net torque on the airplane. Therefore, to be balanced, the center of lift and the center of gravity must coincide. For the example of the symmetric wing in the above section, only the neutrally stable wing is balanced.

Our desire is to have an airplane that is balanced and stable. But if the only balanced wing is neutrally stable, then we cannot have both stability and balance at the same time. This is where the horizontal stabilizer enters the picture.

The Horizontal Stabilizer

Introducing the horizontal stabilizer allows adjustment of the center of lift. If the horizontal stabilizer is lifting upward, it moves the center of lift aft. If it is pulling down, it moves the center of lift forward. So the horizontal stabilizer can be used to balance the airplane. But how does it contribute to stability? Before we can answer this question, you must understand the concept of the *neutral point*.

The example of the wing in the previous section showed stable, neutrally stable, and unstable configurations. This was determined by the location of the center of gravity with respect to the center of lift of the wing. For the symmetric wing described, the center of lift remains fixed and is called the neutral point of the wing. This means that when the center of gravity is located at this point, there is neutral stability.

When we add a horizontal stabilizer, the neutral point is changed due to the lifting ability of the tail. If the center of gravity of the airplane is at the neutral point, the plane has neutral stability. Now let us look at some examples to illustrate the effects of the horizontal stabilizer.

In Figure 4.3 a toy airplane is shown in its trim condition (i.e., straight-and-level flight) and then perturbed with an increased angle of attack. Three examples are illustrated: stable, neutrally stable, and unstable. The top example is chosen such that the center of gravity is ahead of the neutral point. Looking at the picture, we see that the lift of the wing produces a torque that wants to push the nose down since the airplane rotates about its center of gravity. To counter this rotation and thus balance the airplane, the horizontal stabilizer must produce a downward force on the tail of the airplane. Notice how the horizontal stabilizer is at a small negative angle of attack. Because the tail is so far aft, there is a long lever arm so that the downward force on the tail only needs to be on the order of 10 percent or less of that of the wing (typically 6 to 8 percent). Because of the downward force on the tail, the center of lift moves forward from the center of lift of the wing to coincide with the center of gravity. The toy airplane is balanced.

The current $130 to $140 million price tag for a Boeing 777 is roughly the cost of building an entire shopping mall.

The picture on the top right in Figure 4.3 shows the stable toy airplane perturbed from its trim condition. Just as an example let us say a gust has increased the angle of attack by 5 degrees. The wing will create more lift because of the higher angle of attack. Meanwhile the downward force on the tail will be reduced because of a decrease in its negative angle of attack. The net increase in lift on the wing and reduced downward force on the horizontal stabilizer results in the center of lift moving aft. The airplane is no longer balanced but is stable, because there is a restoring torque. So the airplane will pitch back toward the balanced condition, where it will resume straight-and-level and balanced flight.

The toy airplanes in the center of Figure 4.3 illustrate what happens when the center of gravity is exactly at the neutral point. Notice that

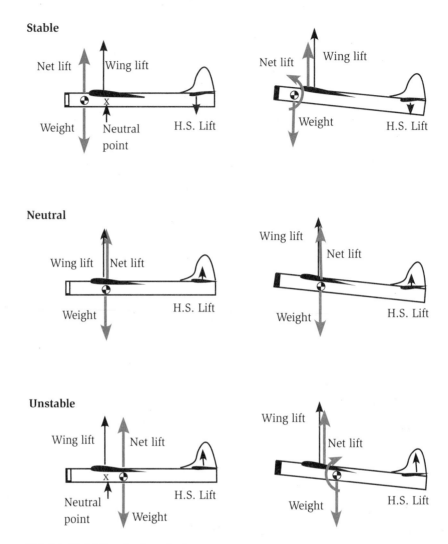

Fig. 4.3. Stability of a toy airplane.

the center of gravity is slightly aft of the center of lift of the wing alone. Thus, to balance, the horizontal stabilizer must produce a small positive lift. In this particular situation, the center of lift is independent of the angle of attack. In other words, when the airplane is perturbed from straight-and-level flight, the increased lift on the wing and the increased lift on the tail balance such that the center of lift remains stationary. This happens at the unique location where the wing's lift and tail's lift change together in exact balance. So, when the airplane pitches up due to some disturbance, as shown with the toy airplane on the right, there is no restoring torque about the center of gravity. This toy airplane is neutrally stable, but balanced.

Finally, the bottom example in Figure 4.3 shows what happens if the center of gravity is behind the neutral point. In straight-and-level flight, as shown on the left, the lift on the wing is balanced by the lift on the tail. But, in its perturbed state, for example, 5 degrees nose pitch up, the lift on the wing grows faster than the lift on the tail. The result is a rotational torque that rotates the toy airplane farther from its initial straight-and-level flight state. This of course is unstable.

One question which you might ask is how the lift on the wing can grow faster than that on the tail since lift is just proportional to the angle of attack. There are two reasons. First, the tail is generally less efficient. On a typical airplane, the tail has about half the aspect ratio of the wing. You learned in Chapter 2 the importance of aspect ratio on efficiency of the wing. The same is true with the horizontal stabilizer. The other factor that affects the horizontal stabilizer's lift is that it is flying in the downwash of the wing. If the wing is generating more lift, it has a greater downwash. The net effect of the downwash is that the horizontal stabilizer sees a lower relative angle of attack than it would if there were no downwash. The wing may experience a 10-degree change in the angle of attack while the horizontal stabilizer only sees a 6-degree change in the direction of the air.

> The accident rate for general-aviation aircraft is a little over 7 per 100,000 hours flown.

The lesson is that for longitudinal (pitch) stability it is crucial that the center of gravity of an airplane be forward of the neutral point of the airplane. That is, the plane is nose-heavy. Such an airplane, when disturbed by a gust of air or a sudden control movement, will tend to return to the original attitude. If the center of gravity is behind the neutral point, the airplane is not flyable. Any disturbance will be magnified and will tend to increase. Thus, if the nose pitches up just a little, the airplane will want to exaggerate this motion. In such a situation the controls will not respond at all. So the airplane must be loaded ahead of the neutral point. That is why children put paper clips on the nose of their paper airplanes.

Pilots must determine the center of gravity before each flight, depending on fuel, passengers, and payload, to ensure that the

The horizontal stabilizer's job is to maintain stability.

airplane's center of gravity is within limits. Unfortunate accidents have resulted from pilots inadvertently taking off with an airplane loaded such that the center of gravity was behind the neutral point. In this situation, no effort by the pilot can save the airplane from catastrophic consequences.

The purpose of the horizontal stabilizer is frequently misunderstood. Frequently, the horizontal stabilizer is said to control pitch. Actually, as discussed in Chapter 1, it is the elevator, which is part of the horizontal stabilizer, that controls pitch. The horizontal stabilizer's job is to maintain stability.

Trim

An airplane must be able to be balanced and stable for a variety of loads. For example, a commercial airplane may face two extremes, one where the airplane is half full of passengers who are all sitting in the forward seats and one where the airplane is half full and all of the passengers are sitting in the aft seats. The two seating arrangements will move the center of gravity of the airplane. We assume that in both cases the center of gravity is ahead of the neutral point so that it is stable. But the airplane must also be balanced. This is where elevator trim plays a role.

Depending on where the center of gravity is, the horizontal stabilizer's lift must be adjusted to balance the airplane. The pilot could achieve this by holding the elevator in position, but this would be tiring on a long flight. So, instead, the pilot trims the airplane to balance it. In Chapter 1 the trim tab was introduced. This small flap on the elevator is adjusted from the cockpit so that the horizontal stabilizer produces the right amount of force to balance the airplane without moving the elevator. On many airplanes, such as the large commercial transports, the entire horizontal stabilizer is rotated so that its angle of attack can be adjusted for different flight conditions.

In 2001, NASA will test active aeroelastic wings that will use wing warping rather than control surfaces. The first to use this method were the Wright brothers, who developed wing warping for roll control before 1903.

Now, what happens when passengers move about in the airplane? The center of gravity moves and the airplane will want to pitch nose up or down. So the horizontal stabilizer trim is adjusted to compensate. But, if the pilot wishes to purposely change the pitch, or angle of attack, the pilot will command the elevators. Thus, the horizontal stabilizer and trim tabs maintain pitch stability and it is the elevator that controls the pitch.

Flying Wings

Today there are examples of airplanes that have no horizontal stabilizers, such as the B-2 bomber shown in Figure 4.4. How can that be? It would seem impossible for the airplane to compensate for any shift in loading. If the B-2 copilot decides to move to the aft cabin, will the B-2 flip? This is the tricky part of designing a flying wing. In this case control surfaces on the trailing edge of the wing move up or down in unison, like an elevator, which moves the center of lift forward or aft, depending on the location of the center of gravity. These control surfaces both balance and stabilize the airplane.

John Northrop was a talented engineer who worked for both Lockheed and Douglas. Northrop was a strong advocate of flying wings. He started his own company to develop the flying wing with the first model, a piston-powered version of the YB-49 (Figure 4.5).

Figure 4.5 shows a picture of the Northrop YB-49 Flying Wing, which flew in the late 1940s. In the picture one can see the two control surfaces, both in the down position, allowing them to act as elevators and flaps.

Fig. 4.4. B-2 bomber.

Fig. 4.5. Northrop YB-49 Flying Wing.

Horizontal Stabilizer Sizing

How big should the horizontal stabilizer be? There are two ways to increase the effectiveness of a horizontal stabilizer. One is to increase the stabilizer's area and the other is to increase its distance from the wing. In fact, it is the distance to the horizontal stabilizer times the wing area of the horizontal stabilizer that dictates stability. Two stabilizer configurations, one with half the area but twice as far from the wing as the other, will have the same stabilizing effect. So the engineer must determine the lever arm, which is the fuselage length, and the horizontal stabilizer area to design a statically stable airplane.

The horizontal stabilizers on the Boeing 777 have approximately the same area as the wings on a Boeing 737.

The neutral point of an airplane is a function of the horizontal stabilizer lever arm times the area. The greater this product, the farther aft the neutral point. The advantage to having a neutral point farther aft is that it gives increased limits on center of gravity location. The increased limits give the airplane greater loading flexibility.

On the other hand, there is a negative impact of increasing either the lever arm or the area of the horizontal stabilizer. Both translate into higher weight, the airplane designer's nemesis. Both also contribute to higher drag due to skin friction. Another problem can be that the airplane becomes too stable. In other words, the effectiveness of the horizontal stabilizer is so great that the pilot has a difficult time changing the pitch angle. This comes under the category of handling properties, which will be discussed later. So sizing a horizontal stabilizer involves a combination of versatility in center of gravity, weight, drag, and handling properties.

Directional Stability

In the previous section we discussed only stability in pitch, known as longitudinal stability. In Chapter 1 you were introduced to two other axes, roll and yaw. Roll stability, known as *lateral stability,* was covered in detail in Chapter 3, on "Wings." The effects of dihedral and sweep were presented and will not be repeated here. Directional stability is the stability in the yaw axis, and gives rise to the vertical stabilizer. The vertical stabilizer and rudder serve the same function as the horizontal stabilizer and elevator, except in yaw, instead of pitch. The rudder is used for control and the vertical stabilizer is for stability. The main function of the vertical stabilizer is to help the airplane *weathervane* and keep the nose pointed into the direction of flight.

The desire for directional stability is to have the airplane always line itself with the wind. So, if a gust temporarily perturbs the direction the nose is pointed, the tail will have a nonzero angle of attack with the airflow, as shown in Figure 4.6. This causes a restoring force to realign the tail with the direction of travel. The effects of misalignment with the flight path are primarily high drag and poor turn coordination.

The Spirit of St. Louis took less than 2 months to design and build.

The size of the vertical stabilizer depends on several factors. For a single-engine airplane, the requirement that sets the minimum size for the vertical stabilizer is that the vertical area of the airplane aft of the center of gravity be larger than the vertical area forward of the center of gravity. This is the same requirement that puts feathers on arrows for stability. A larger vertical stabilizer is needed to counter propeller rotation effects and adverse yaw in a turn, which was discussed in Chapter 3. A single-engine airplane can get away with the minimum-size vertical stabilizer but will require more work on the pilot's part.

For multiengine airplanes the size of the tail is dictated by the torque caused by the loss of one engine. The net thrust being off center

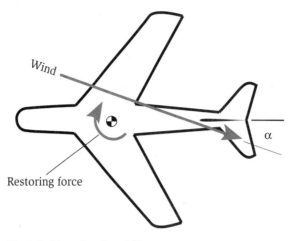

Fig. 4.6. Directional stability.

causes the airplane to want to yaw. A large vertical stabilizer, with trim, can compensate for this. That is why twin-engine commercial transports have such large vertical stabilizers.

The FAA dictates limits on directional stability. Modern airplanes now have vertical stabilizers that are so effective as to make the use of the rudder for small corrections almost unnecessary.

The B-2 bomber, which has no vertical stabilizer, accomplishes directional stability by using differential drag. The ailerons on the side you wish to turn will split, deploying both up and down. Thus, they increase the drag on that side, pulling the nose around. This can be seen on the right wing in Figure 4.4.

Dynamic Stability

Static stability deals with the tendency of the airplane when disturbed to return to its original flight attitude. Dynamic stability deals with how the motion caused by a disturbance changes with time. The three types of dynamic stability are shown for the longitudinal stability of an airplane in Figure 4.7. The first flight path shows positive dynamic stability. When the airplane pitched up, there was a restoring force (statically stable). The path oscillates through the original altitude and

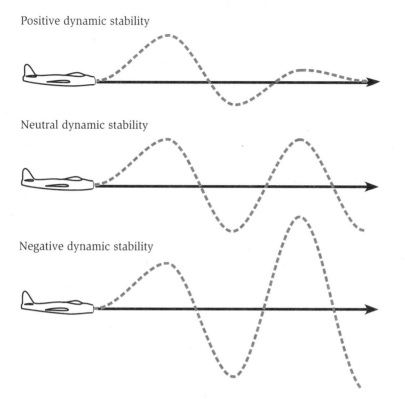

Positive dynamic stability

Neutral dynamic stability

Negative dynamic stability

Fig. 4.7. Three types of dynamic stability.

with the oscillations decreasing with time. This is like a car with good shock absorbers.

The second case is neutral dynamic stability. The airplane is statically stable because there is a restoring force. But the amplitude of the oscillations in this case does not decrease with time. This is kind of like a car without shock absorbers that hits a bump.

The third case in the figure is negative dynamic stability. Again the airplane is statically stable but the amplitude of the oscillations increases with time. This is kind of like a car without shock absorbers going down a "washboard" (bumpy) road.

As with static stability, dynamic stability must be considered in the longitudinal, lateral, and directional (pitch, roll, and yaw) axes. What can make dynamic stability even more interesting is that more than one axis can couple, producing some very interesting motion.

The French Voisin brothers founded the first airplane building company in 1906, despite the fact they had never successfully flown.

In the following sections you will be introduced to three modes of dynamic motion that are the easiest to understand and may be most familiar to you.

Phugoid Motion

Have you ever thrown a paper airplane and watched it follow a flight path that climbs and slows and then descends and speeds up, as illustrated in Figure 4.8? This type of motion is common to all aircraft and is given the name *phugoid motion.* Phugoid motion is a trade between kinetic and potential energy, that is, speed and altitude. It occurs at a constant angle of attack so as the speed increases, so does the lift. The extra lift causes the airplane to increase altitude. As it does, the airspeed falls off, decreasing lift, and thus eventually altitude. In the extreme, at its maximum height the airplane will lose so much speed that it will stall.

But there is no need to worry about this in a real airplane. The time it takes to complete one cycle, the period, is on the order of minutes. Only a sleeping pilot would ignore this. In fact, the period is so long, most pilots do not even recognize that they are controlling this motion. Also, the oscillations will damp out eventually by themselves if they are ignored.

Dutch Roll

The *Dutch roll* is a motion that couples roll and yaw. The name comes from the motion of the Dutch speed skaters as they glide across the ice. It is kind of like the coupling of a small rolling motion with a small

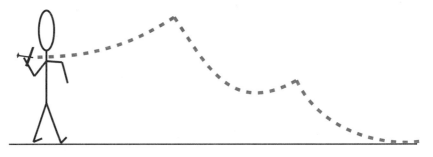

Fig. 4.8. Phugoid motion.

wiggle of the tail. The airplane does this while maintaining its heading. Dutch roll has a short period and presents no major stability problem. The biggest problem is upset stomachs of passengers in the rear of the airplane. Because of passenger discomfort, commercial airplanes use *yaw dampers* which move the rudder to automatically damp out Dutch roll. A yaw damper is not a critical flight control, so a flight will continue if the yaw damper is inoperable. But there may be more sick passengers on the flight.

The first Academy Award for Best Picture was awarded to <u>Wings</u>.

In smaller airplanes most pilots do not feel the motion of Dutch roll. That is because all motion occurs around the center of gravity. In a small airplane, the pilot and passengers are usually sitting very close to the center of gravity. Only on a large commercial transport do passengers find themselves far enough away from the center of gravity to feel this motion.

Spiral Instability

The last instability that we will consider is in spiral motion, frequently referred to as *spiral divergence.* This is instability in yaw and roll, which leads to a downward spiral. An airplane that has a spiral instability will eventually spin into the ground if the pilot does not intervene. However, although the results are dramatic, the FAA allows for some spiral divergence in airplane designs. This is because, like phugoid motion, spiral divergence takes a long time to develop. Again, only a sleeping pilot could ignore this. Note that Lindbergh's unplanned nap might have been disastrous if he had not woken up on time. The *Spirit of St. Louis* was fairly deep into a spiral divergence at the time he woke up.

Stability Augmentation

The newest military aircraft are exploring statically unstable airplanes, which could not be flown without computer control. What the computer does is kind of like when you try to balance a pencil vertically on the tip of your finger. It is very difficult. This is a statically unstable situation. But, if you could react quickly enough, you could keep the pencil in position. Quick reaction is the role of the computer in the unstable airplane.

Why would the military want a statically unstable airplane? As discussed in Chapter 3 on "Wings," the answer is maneuverability. If the airplane has a natural tendency to diverge from a specific condition, such as straight-and-level flight, then it will be much more responsive when the pilot wants to make a change. Another reason for designing a statically unstable airplane is that smaller stabilizers might be used, which decreases the weight and drag of the airplane.

To remain stable, the flight control system on the X-29 (Figure 3.8) has to make 40 corrections per second.

The ability of a computer to solve problems quickly gives it a tremendous advantage. The Wright brothers preferred their airplanes slightly unstable so that the pilot would have to react and pay attention. But the design was only slightly unstable, and controllable by use of the control surfaces. If an airplane were highly unstable, the pilot would not be able to react fast enough to compensate. But a computer might be able to react fast enough. Today, inserting a computer into the control loop can augment stability. The pilot can manage the overall flight path while the computer manages the quick response tasks, or the computer can do both.

With the computer in the control loop an airplane can be built to be naturally unstable. One of the first examples of an unstable design was the X-29, shown in Figure 3.8. The computer makes fast, tiny adjustments that allow the pilot to focus on other tasks. If the computer were to fail, the airplane would instantly become uncontrollable to the pilot, with fatal consequences.

Handling

As mentioned earlier, handling qualities were not quantified until after WWII. Before that time handling qualities relied on pilot opinion. Words like *hot, fun, smooth, fast, sluggish,* and *sporty* are still used by pilots to qualify an airplane's handling properties. But what is *sporty* to one pilot might be *sluggish* to another. The more dangerous situation is the opposite, where the high-time fighter pilot, who considers the airplane to be *smooth,* turns the airplane over to a novice, who finds it rather *sporty.* This is a particular problem in the home-built industry, which rarely publishes qualitative handling data.

One handling quality is *stick force.* This is a measure of how much force is required to make a certain change in a control surface. Suppose an airplane required 40 lb of force to roll the airplane at 1 degree per second. This would qualify as *extremely sluggish.* But, if a 1-lb force on the controls corresponds to a 180-degrees-per-second roll rate, this would be *very sporty.*

Another issue is *control balance.* Suppose you have to put 5 lb on the control yoke for maximum roll but 30 lb for maximum pitch. This is an unbalanced control system. Ideally, 5 lb on the control yoke should give roughly the same changes in both the roll and the pitch axis.

Another factor is the *adverse yaw.* Older airplanes had significant adverse yaw, so a pilot had to be diligent with rudder pedals. A modern trainer hardly needs any rudder input to counter adverse yaw. The improvement has come primarily through the use of dihedral and larger vertical stabilizers. Older pilots consider this sloppy flying, but the realities of a modern trainer are that rudder pedals are barely needed.

> The Wright brothers did not fly from October 16, 1905, to May 6, 1908, to protect their pending patent.

Fly-by-Wire

Before the days of the digital computer, airplane control surfaces were linked to the control yoke through cables, push rods, and hydraulic lines. These were mechanical links from the pilot's controls to the control surface. In the case of cables and push rods, the stick force was a matter of designed mechanical advantage that is utilizing the basic concept of a lever. The problem with cables and push rods is that they can be difficult to route from the yoke to the control surface. For example, a cable or push rod would be undesirable down the center of the cabin. Hydraulic lines, on the other hand, can be routed fairly easily, since they are just tubing.

> Jimmy Doolittle, famed air racer with a Ph.D. from MIT, was the first to fly coast to coast in the United States on Sept. 4, 1922.

With the computer in the loop there is no need for direct mechanical connections or running hydraulic lines through the airplane. A fly-by-wire system is a control system where control actions are transmitted by wire. The pilot inputs a command on the yoke, which is read by a computer. The computer translates the command, along with its own inputs to augment stability, to an electrical signal. A wire then connects the cockpit to various *actuators,* which convert the signal into a mechanical action, like moving the elevator.

There are a few interesting side effects of fly-by-wire. One is that an intelligent computer can be used to make decisions. For example, the computer may monitor the angle of attack and not allow the aircraft to reach the stall angle of attack. Thus, no matter how hard the pilot pulls back on the control yoke, the airplane will not increase its angle of attack to a stall. This can be useful in a fighter aircraft where the pilot in combat does not have time to watch the angle of attack indicator. Another example is that the computer might turn a sloppy, pilot-controlled landing into a smooth landing. In essence, the pilot and the computer both fly the airplane.

An unfortunate incident happened at a French air show in 1988. An Airbus A-320 on a demonstration flight crashed off the end of the runway. The jet was scheduled to do a flyby. But the computer interpreted the approach as an approach to landing. The subsequent confusion between the pilot and the computer resulted in neither a controlled flyby nor a controlled landing. Instead, the airplane crashed and was destroyed. The A-320 was the first civilian airplane to use fly-by-wire control.

In Sioux City, Iowa, on July 19, 1989, a DC-10 landed after losing all tail surface controls due to an explosion in the middle tail engine. The pilot miraculously maneuvered the airplane to a crash landing using the thrust from the two wing engines to turn.

Another side effect worth noting is the effect on stick force. Flying an airplane with a joystick is no different from flying in a computer simulation with a joystick. The joystick on the computer has no way of giving mechanical feedback in terms of resistance to turning. So it takes the same force to make a tight turn at low speed as at high speed. This feedback is a point of contention between pilots and designers. The pilots want the mechanical feedback. Now at least one airplane manufacturer has developed springs and linkages that are controlled by the computer system that give the pilots the same feel as the old mechanical linkages. Of course, now this means that the stick force and balance can be tuned with software. But, at the same time, this is extra hardware, and thus weight, which only serves the purpose of making the pilot more comfortable. However, safety in itself provides a justifiable argument in favor of such systems.

The computer has resulted in an interesting aspect of handling properties. With the computer installed between the pilot and the control surfaces the properties of stick force and balance can be changed in software. This can be useful in a new airplane development program. For example, the expected handling properties of the Boeing 777 were programmed into a Boeing 757, which was flown extensively by the test pilots before the first Boeing 777 was even built. The first flight of the Boeing 777 went off without a hitch, in part because the pilot had many simulated hours flying a Boeing 777.

> On June 10, 1965, the first fully automatic landing of an in-service commercial airplane occurred at London, Heathrow airport.

Wrapping It Up

In this chapter you were introduced to basic concepts of stability and control. There is a great deal more detail that could be covered on this topic, but you have been given a basic introduction as a starting point.

In the early days, stability and control were not as significant a part of the airplane's design as they are today. On a modern airplane, commercial and military, the stability and control systems include computers. The computer has many capabilities, and engineers are learning more ways to use them. So, in addition to augmenting stability, an airplane's computer may take navigation data, instrument landing data, air traffic data, and weather data and create and execute a flight path. In fact, the computer alone can fly an entire flight, from takeoff to landing. The pilot's job becomes more of a systems monitor than the person flying an airplane. There is a joke that a pilot and a dog will fly the airplanes of tomorrow. The dog's job will be to bite the pilot's hand if he touches the controls and the pilot's job will be to feed the dog.

> The cost of modern airplane computer and electronics systems is approaching 50 percent of the cost of the airplane.

CHAPTER 5

Airplane Propulsion

The propulsion system is one of the most complex systems on an airplane, yet the principles behind airplane propulsion are not very complicated. An airplane in flight requires power to provide lift and to overcome the drag associated with the impact of the air on the airplane. If it is climbing or making a turn, additional power is needed. In Chapter 2, "How Airplanes Fly," we showed that in producing lift, work is done on the surrounding air. Here we use similar arguments to describe how aircraft propulsion systems work.

The power required for climbing and turning is discussed in Chapter 7, "Airplane Performance." In this chapter, you will learn how propulsion systems provide the necessary power for flight. You will also learn how jet engines differ from piston engines in how they deliver the needed power. Some of the differences, and similarities, may surprise you.

With some exceptions, the gliders being the most notable, the power needed for flight is provided by either a piston or a jet engine. The energy produced in these engines must be transferred to the surrounding environment to propel the aircraft. These propulsion systems require some very complex engineering. But one need not understand the details of those systems to understand what the propulsion system is doing and why engines look the way they do.

It's Newton Again

You learned in Chapter 2 that for a wing to create lift it must divert air down. This principle is exactly how an aircraft propulsion system works, except that to create thrust it must push air back. Just as your household fan pushes air back, so does a propeller and a jet engine. Fortunately, your house fan does not have enough thrust to propel itself.

Also like wings, aircraft propulsion systems are applications of Newton's laws. Remember Newton's third law states that "for every action, there is an equal and opposite reaction." In an aircraft propulsion system, the action is acceleration of air or exhaust and the reaction is the force, or thrust, produced. Again we use the alternate form of Newton's second law which states that the thrust is proportional to the amount of gas accelerated per time times the speed of that air.

> Harriet Quimby was the first U.S. woman to earn a pilot certificate and in 1911 was the first woman to cross the English Channel.

Thrust

Though it is not an aircraft propulsion system as such, studying the rocket engine is an excellent way to understand propulsion. An example of how a rocket engine works is shown in Figure 5.1. Fuel and an oxidizer are pumped into a combustion chamber, producing a large amount of gas at a high pressure. The gas accelerates to the throat of the motor, where it reaches a velocity of Mach 1. After the throat the

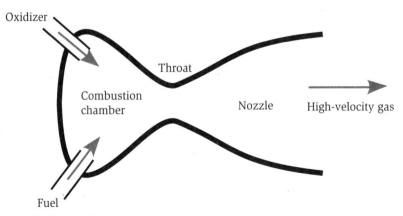

Fig. 5.1. Schematic of a rocket engine.

gas continues to accelerate, producing an exhaust of hyper-sonic gas with a great deal of thrust. A large rocket expels a great deal of gas at a high velocity. The magnitude of the amount of gas expelled can be seen in Figure 5.2, which shows the launch of Apollo 8.

The thrust from a rocket motor is analogous to the recoil of a rifle when a bullet is fired. In order to increase the thrust of a rocket, one can increase the amount of gas expelled per second, the velocity of that gas, or both. An aircraft propulsion system works much the same way as the rocket motor. The force on the airplane is a reaction

The Apollo IV was the noisiest of all rockets. It was so loud it produced seismic readings in New Jersey, a distance of 850 mi (1400 km).

Fig. 5.2. Apollo 8 launch. (*Photo courtesy of NASA.*)

to accelerating air or exhaust. Newton's second law states that the thrust is equal to the amount of mass, per time, pushed by the engine (propeller or jet) times the increase in velocity the air experiences. If you have ever stood behind an airplane with its propeller turning, you can certainly recognize that a great deal of air is being blown back.

Power

Aircraft propulsion consists of two distinct parts. There is the engine that converts a source of energy, such as fuel, to work. Then there is the part of the system that converts the work of the engine into work on the surrounding environment to produce propulsion. The most obvious example is a propeller. A piston engine and propeller combination is an example of a complete aircraft propulsion system. A turbojet is another example, but the parts of that system are a little harder to distinguish from each other.

Usually engineers, flight instructors, and educators relate flight and propulsion in terms of forces. In this book, we take the perspective of power, which is adjusted by the throttle and can be measured by the pilot. Looking at propulsion from power carries some intuition with it. If one increases the throttle or fuel flow, the power increases. Power is the rate of using energy, or doing work, which is the key to understanding propulsion. Power also lends itself to another fundamental concept: efficiency.

Looking at the propulsion system from the standpoint of power, it is convenient to introduce a few terms. You know that power is required for flight: for supporting the weight of the airplane, for climbing, for turning or accelerating. This is the *required power* for flight. The power that is actually produced by the engine and delivered to the propeller or is available for propulsion by the jet we will call the *engine power*.

The rocket-powered Bell X-1 was the first airplane to go supersonic.

The power actually used to produce thrust is the *propulsive power*, which is just equal to the thrust produced times the speed of the airplane. The propulsive power is always smaller than the engine power because of inefficiencies. The difference between the engine power and the propulsive power we will call called the *wasted power*.

This is the power that is lost to kinetic energy in the propeller's slipstream or the jet's exhaust.

Again, the power delivered by the engine is the engine power, and the power that goes into propelling the airplane is the propulsive power. With a piston engine, for a fixed engine power, the propulsive power depends only slightly on the airplane's speed. The first figure in Figure 5.3 illustrates how the thrust and propulsive power vary with speed for a typical piston engine/propeller combination at a fixed power setting. Since thrust is power divided by speed, a fairly constant propulsive power means that the thrust provided by the propeller decreases with speed. The thrust from a propeller degrades with speed but the propulsive power holds up quite well. Of course, the details are dependent on propeller design, among other things.

As can be seen in the second figure, the jet engine behaves quite differently with respect to speed than a piston engine. The thrust available to a jet engine is roughly constant with speed and the propulsive power is proportional to the speed of the airplane. This will be discussed in more detail later, though one can anticipate that the differences will impact the performance of an airplane.

> The lifetime cost of a Boeing 747-400 per seat-mile is the same for an automobile.

Efficiency

The objective of an aircraft propulsion system is to produce the required power as efficiently as possible. There are basically two areas where propulsion systems lose efficiency. The first is in the conversion

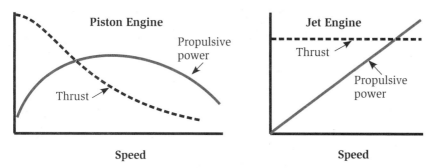

Fig. 5.3. Propulsive power and thrust as functions of speed for a propeller and a jet engine.

of fuel to engine power. This is the *engine efficiency*. The losses here are primarily due to inefficiencies in the burning of the fuel and friction within the engine. Energy is also used for engine support such as powering fuel and lubricant pumps and the generation of electricity. These losses reduce the engine's output efficiency.

Once the engine has converted the chemical energy of the fuel into mechanical energy, it must convert the mechanical energy into propulsion. The difference between the engine power and propulsive power gives us the *propulsive efficiency*. The total efficiency of the propulsion system, which is engine efficiency times propulsive efficiency, is a measure of how much power the system develops for a certain quantity of fuel burned. So, what contributes to the propulsive efficiency? The same arguments used for lift efficiency can be used here.

> Because of the oil embargo, the price of 1000 gallons of Jet Al fuel went from $100 in 1970 to $1100 in 1980.

Remember from Chapter 2 that the lift of a wing is proportional to the momentum mv that is transferred to the air per time. The kinetic energy given to the air by the wing ($\frac{1}{2}mv^2$) is an energy loss to the airplane. For the most efficient flight one wants to produce the necessary lift while transferring as little energy to the air as possible. Therefore, one wants to divert as much air as possible at as low a velocity as possible. That is why the efficiency of a wing increases with size. Increased lift efficiency requires the increase in the amount of air diverted, not an increase in the speed of the diverted air.

Propulsion systems produce thrust and deliver their power to the surrounding environment in the same manner as a wing. Thus, for the most efficient thrust one wants the engine to accelerate as much air as possible at as low a velocity as possible. This minimizes the wasted power. If a propeller or a jet could discharge a very large amount of air or exhaust at a relatively low velocity, it would take much less power than another system that developed the same thrust but by discharging a small amount of air at a high velocity. Keep in mind that the energy given to the air producing the thrust is lost energy that must be paid for by the engine.

> Air transportation consumes 8.9 percent of the world's transportation energy budget. Highway transportation consumes 72.4 percent.

The propulsive power divided by the engine power is the propulsive efficiency. For the best propulsive efficiency you would want to have almost zero kinetic energy in the air left behind. In this case the wasted power would be almost zero and the engine power would equal the propulsive power, which is the thrust times the speed of the airplane. Unfortunately this is not possible. To produce thrust, there must always be some velocity given to the air and thus some waste.

Let us look at the situation of an engineer who wishes to design a jet engine that is to have twice the thrust of a previous engine. The engineer can increase the thrust by increasing the amount of mass discharged by the engine per time, by increasing the velocity of the exhaust, or by increasing both. However, if one increases the exhaust velocity, the wasted power increases as the velocity squared. So doubling the thrust by doubling the exhaust velocity doubles the propulsive power but increases the wasted power by a factor of 4! However, if one doubles the thrust by doubling the mass discharged, the propulsive power has again been doubled and the wasted power has only doubled. Basically, the objective of an aircraft propulsion system is to create the most thrust for the least wasted power, which leads aircraft engine design to favor increasing mass flow over exhaust velocity.

> Each engine of the Boeing 777 produces 98,000 lb (436,000 N) thrust and has an airflow of 1.4 tons (1.3 metric tons) of air per second.

Even for the most efficient propulsion systems, a great deal of kinetic energy is imparted to the air. There is no ideal propulsion system for flight. Wings are much more efficient in producing lift than engines in producing thrust. This is because wings, with their large size, are able to divert a great deal of air at a slow speed. Engines do not have that luxury.

Propellers

A propeller is simply a rotating wing. As an illustration of this point, look at Figure 5.4, which is a photo of slices of a propeller. The airfoils of the propeller are clearly winglike. We know that the purpose of the wing is to divert air down to create lift. Likewise a propeller diverts air back to produce forward thrust. For low-speed flight a propeller is the most efficient means of propulsion. At peak efficiency 85 percent or more of the engine power can be converted into propulsive power by a propeller. Thus only about 15 percent of the power is wasted.

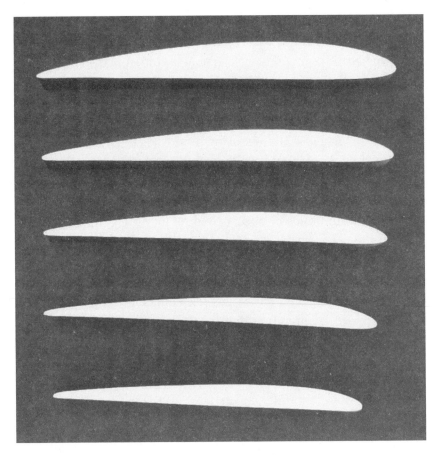

Fig. 5.4. Photos of propeller slices. (*Photo courtesy of Albert Dyer.*)

The world indoor FF (Free Flight) record as of January 1, 2000, was 60 minutes and 1 second. On June 1, 1997, this was achieved by Steve Brown's Time Traveler. Free Flight is a competition using rubber-band-powered airplanes. This same airplane flew for an unofficial 63 minutes and 54 seconds on another flight.

Since the propeller works by pushing a certain amount of air mass through each revolution and accelerates this mass of air to a higher velocity, we can begin to understand the tradeoffs in propeller design. A large-area propeller will be more efficient than a smaller one because it can push more air. To obtain the same thrust, the smaller propeller will accelerate less air but to a higher velocity, making it less efficient. Rubber-band-powered model airplane enthusiasts know that a large, slow-turning propeller will result in the longest flights.

Propeller size and rotation speeds are dictated by many factors. First, a large, slow-turning propeller may not be practical because of ground clearance. But of equal importance is matching the propeller rotation speed to the type of engine provided. Couple these requirements with the need to keep the propeller-tip speed below the speed of sound (for reasons of noise and additional power loss) and you have propellers that we see on airplanes today. Airplane piston engines are usually designed to operate at between 2200 and 2600 rpm (revolutions per minute). Limiting the tip speed to roughly the speed of sound in normal flight situations gives a typical propeller diameter of 72 to 76 inches (182 to 193 cm).

The power added to the air is proportional to the propeller's rotational speed cubed. This can be easily understood by considering a wing with a fixed angle of attack. We know, from Chapter 2, the power imparted to the air by the wing is proportional to the amount of air diverted times the vertical velocity of that air squared. If the speed of the wing is doubled, while keeping the angle of attack constant, both the amount of air diverted and the vertical velocity of that air will also be doubled. Thus the power will have gone up by a factor of 8. Likewise with a propeller, with a fixed pitch and constant forward speed, the power transferred to the air is proportional to its rotational speed cubed. What this means is that the power needed to turn a propeller increases very rapidly with its rotation speed. It is therefore very important to get the area of the propeller correct for the engine size. If the blade is too small, the load on the engine will be low and the engine will "over-rev" by going to a high rpm. This can damage an engine. If the blade is too large, the engine will not be able to reach its optimum operating speed and thus will not be able to deliver full power to the propeller.

The radial engines in WWI fighter airplanes actually rotated with the propeller.

Multibladed Propellers

The total area of the blades on a propeller determines the ability of the propeller to convert the engine's power into thrust. The more area, the more power the propeller can convert. For most small planes the appropriate blade area is achieved with two blades, with the area of the blades becoming greater with increasing engine size. Some propellers increase the total area by using three-bladed, four-bladed, and even six-bladed propellers. The reality is that if the total area is the same for two-bladed vs. three-, four-, five-, or six-bladed propellers, the efficiency will be close to the same.

Transitioning from two to more blades with the same total area is a result of subtle tradeoffs. Two-bladed propellers are usually best for lower-speed airplanes where the power requirements are low. More blades are used when power requirements are higher, such as faster climb and higher speeds. Other factors favoring multiblade propellers are that they produce less objectionable noise and reduced vibrations. So, why not always use more blades? The simple reason is that they are more expensive.

> Wilbur Wright died of typhoid fever in 1912.

Propeller Pitch

With the size and speed of the propeller fixed, the airflow through the propeller is basically fixed. So, in order to get more thrust, the air velocity behind the propeller must be increased. The *pitch* of a propeller is analogous to the wing's angle of attack. With a *fixed-pitch* propeller, the angle of the blades is fixed with respect to its rotation direction. The propeller's apparent angle of attack is determined by the pitch of the propeller, its speed of rotation, and the speed of the airplane through the air. The faster the airplane is traveling the smaller the apparent angle of attack. This is demonstrated in Figure 5.5. As the airplane flies faster, the wind due to the forward speed reduces the angle of attack of the propeller. Thus, the propeller diverts less air, producing less thrust and requiring less power from the engine.

The efficiency of a fixed-pitch propeller depends on the speed of rotation and the speed of the airplane. Figure 5.6 shows the efficiency of a propeller at a single rotation speed for various pitch angles as a function of the speed of the airplane. From this it is clear that for a single pitch, the efficiency is the optimum over a fairly narrow range of airplane speeds. Because of this, a fixed-pitch propeller must have a fairly high pitch for all-around performance. The rated engine power available at a given altitude is determined by the engine's rpm, which is also the propeller's rotation speed. Thus high pitch may cause the engine to run below its optimum speed during takeoff and thus not produce full power. The same blade at cruise speed may require that the engine be throttled back to prevent the engine from operating at

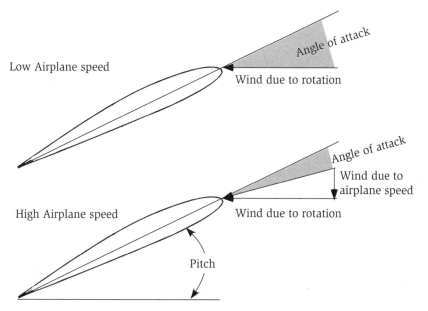

Fig. 5.5. Angle of attack of a rotating propeller.

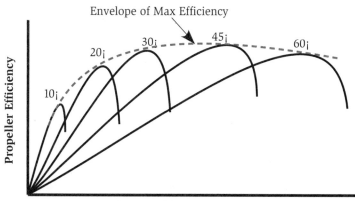

Fig. 5.6. Efficiency of a propeller at a single rotation speed for various pitch angles as a function of the speed of the airplane.

too high an rpm. Thus the fixed-pitch propeller is at best a compromise.

A solution to this compromise is the *constant-speed propeller*. The constant-speed propeller allows the pilot to control both rotation speed and propeller pitch. A constant-speed propeller works something like a governor on the engine. There are two controls, the engine throttle and an rpm control. The throttle controls the power output of the engine and the rpm control sets the rotation speed of the propeller and thus the speed of the engine. If the engine wants to run too fast, the pitch of the propeller is automatically increased until the engine slows down to the preset speed. This allows the efficiency of the constant-speed propeller to look something like the *envelope of maximum efficiency* shown in Figure 5.6.

On takeoff and in climb the propeller is adjusted to have a fairly small pitch. Because of the airplane's slow speed, the angle of attack is still quite large. In cruise the airplane's speed causes the propeller to see a reduced angle of attack. Here the pitch of the propeller is increased to allow the engine to operate at its optimum performance. Typical constant-speed propellers on small airplanes improve the overall efficiency of the propeller by about 15 percent.

> In 1933, Boeing made an exclusive agreement with United Airlines for its advanced model 247. Competing airlines, which could not buy the Boeing 247, asked Douglas to build a competing aircraft. Douglas's answer was the DC-1, which led to the DC-3 by 1936. The Boeing 247 became instantly obsolete.

Piston Engines

> The Wright brothers' first engine produced 12 hp and weighed 180 lb.

Either a piston or turbine engine can power a propeller. Here you will be briefly introduced to aircraft piston engines. We will not go into detail about how they work, but rather focus on the relationship of the piston engine to the power delivered.

In brief, the piston engine works by converting a certain amount of energy from a fuel and oxygen mixture into kinetic energy of a piston. The energy of the piston is then used to turn a shaft. Finally, the shaft turns a propeller. This is illustrated in Figure 5.7.

If we define the total power as the amount of energy per second available in the fuel and air mixture, then the total power will be limited by the amount of air that can be pumped into the cylinders of a piston engine. Larger cylinders mean a more powerful engine. The higher the altitude, and thus the lower the air density, the less power is produced. Because the available oxygen decreases with altitude, the *normally aspirated* piston engine generally limits aircraft to low operating altitudes.

> Modern aircraft piston engines weigh approximately 2 pounds per horsepower.

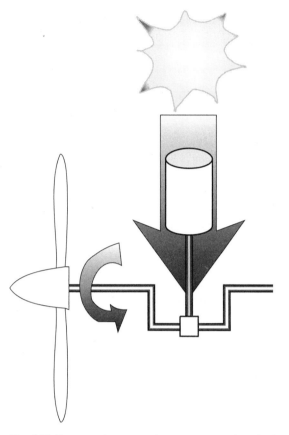

Fig. 5.7. How a piston engine converts chemical energy to propeller rotation.

   One way to overcome the power loss with altitude is to add a pump to the air intakes to increase the amount of air in the cylinders. There are two common methods for doing this. The first is called *turbocharging*. A turbocharger makes use of the energy expelled in the exhaust to run a small pump in the air intakes. A "supercharger" is another method to pump additional air into the cylinders. A supercharger can be powered mechanically through a belt on the engine shaft, or with an electric motor. The purpose of both is the same, to increase the amount of air (oxygen) in the cylinders at higher altitudes where the air is less dense. The result is that a turbocharged or supercharged engine can maintain constant power up to a higher altitude. Above that altitude the pump can no longer maintain sea-level

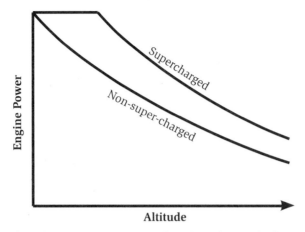

Fig. 5.8. Engine power as a function of altitude for a non-turbo-charged and a turbocharged engine.

density in the cylinders and the power drops off with further increase in altitude. Figure 5.8 shows the engine power as a function of altitude for a non-turbo-charged engine and a turbocharged engine. Turbochargers are usually not used to increase power above the maximum rated power at sea level. Engine temperatures and loads would be too great, resulting in damage to the engine.

The thing to keep in mind with piston-engine-powered airplanes is that the engine's power is a function of altitude but not of the speed of the airplane. As shown in Figure 5.3, thrust of the propeller decreases with speed but the propulsive power is pretty much constant until high speeds are reached.

On Nov. 12, 1906, Brazilian Alberto Santos-Dumont made the first sustained powered flight in Europe. He was hailed as the first to fly, since the Wright brothers' success was still unknown.

The Turbine Engine

So, how is a turbine engine, which is the heart of all jet engines, different from a piston engine? For one thing, it is a little harder to separate the "engine" from the device that produces thrust. So, we start this discussion by introducing the basic elements of the turbine engine in what is called the *engine core*. An important result you will learn is that a turbine works such that the available power increases with speed and the thrust is independent of speed. For a piston engine/propeller propulsion system the available power is constant with speed and the thrust decreases with speed. This is an important difference between the two types of engines and impacts the way the airplanes are flown.

In a turbine engine the energy of the combustion is transferred to the exhaust, rather than a mechanical piston. Figure 5.9 shows how one might conceptualize a turbine engine, starting from the rocket motor in Figure 5.1. The rocket carries fuel and its own source of oxygen. That is how it is able to operate in the vacuum of space. But a turbine engine flies in the atmosphere where oxygen is plentiful. So the rocket motor could be fitted with a compressor to supply high-pressure air to the combustion chamber. To run the compressor, a turbine fan has been placed in the exhaust to convert some of the energy from the high-speed exhaust into mechanical work.

Conceptually, Figure 5.9 has the components of a jet engine, though the implementation of the components is quite different in practice. Figure 5.10 shows a more realistic drawing of a jet engine. Basically it consists of a tube with an inlet (or a diffuser), followed by a compressor for the air, a burner where the high-pressure air and fuel are burned, and a turbine to power the compressor. At the exhaust end of the turbine there is a nozzle to direct the exhaust to give thrust. The three components, compressor, burner, and turbine, are the core of the jet engine.

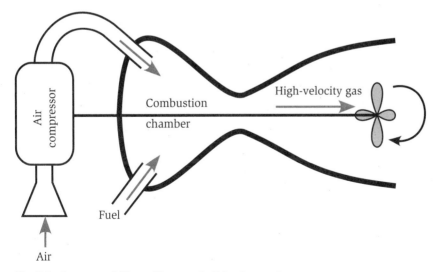

Fig. 5.9. Conceptual idea of how to build a jet engine.

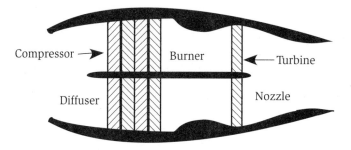

Fig. 5.10. More realistic drawing of a jet engine.

Let us go through the different parts of the turbine engine one at a time. The diffuser and nozzle will be discussed a little later, since they are really not part of the turbine engine itself but additional parts used to make the turbine engine a jet engine.

Compressors

The *compressor* has two functions in the turbine engine. The first is to act like a one-way valve that prevents the combustion gases from blowing out the front of the engine. The compressor's job is to always push air in one direction, into the burner. The other function implied by its name is to increase the pressure and density of the air, and thus the oxygen, so the fuel will burn efficiently. This is particularly important at high altitudes where there is very little air.

There are two types of compressors used in jet engines. Larger engines use *axial-flow compressors* as shown in Figure 5.11. As the name implies, the air flows down the axis of the compressor, where it is fed directly into the burner. Smaller jet engines are more likely to use *centrifugal compressors,* or impellers, as shown in Figure 5.12. This type of compressor "pumps" the air to the outer radius of the engine, where it is then redirected into the burner. Though different in construction, the purpose of both compressors is the same: to compress the air. We now look at the two types in more detail.

AXIAL-FLOW COMPRESSORS

In the *axial-flow compressor* a series of rotating blades pushes the air back and in doing so adds energy to that air. These are called axial-

Fig. 5.11. Axial-flow compressor.

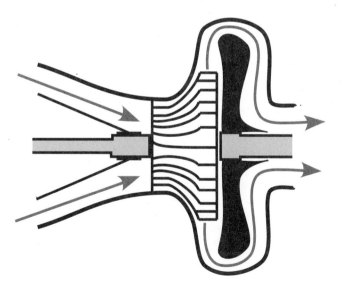

Fig. 5.12. The centrifugal compressor.

> Today, commercial airplanes have to turn back more often because of inoperable toilets than because of an engine failure.

flow compressors because the air continually moves along the axis of the engine. Each blade is basically a rotating wing, just like a propeller. There is one fundamental difference between the compressor and the propeller. With the compressor, the blades are in a duct and therefore the added energy results in an increase in the pressure of the air rather than an increase in the speed of the air. How pressure is produced, rather than speed, is kind of interesting.

The axial compressor is made of rows of blades, made up of rotating blades followed by stationary blades as shown in Figure 5.13. A typical row of rotating blades has 30 to 40 blades and is called a *rotor*. Following each rotor is a stationary set of blades, called a *stator*. The rotor's job is to increase the energy of the air and thus its pressure. The stator increases the pressure of the air further by slowing it down from the speed at which it left the rotor. The pressure increase across a single stage of rotor/stator is fairly low, but multiple stages can produce fairly high compressions with high efficiency.

It is not wise to try to increase the pressure too much across a single stage because this increases the chances of the blades stalling just like a wing that is trying to produce too much lift. The stall causes the flow to reverse and is referred to as *compressor stall*. Rather than trying to increase the pressure substantially across each stage, multiple stages are used to decrease the pressure gain across each stage. The result is that an entire compressor section of 10 to 12 stages may increase the pressure by a factor of 10 or more.

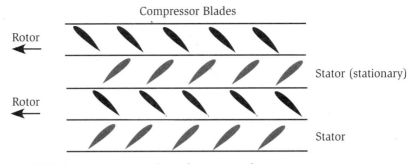

Fig. 5.13. A compressor consists of a rotor and a stator.

CENTRIFUGAL COMPRESSORS

Centrifugal compressors push the air out radially rather than along the axis of the engine. An *impeller,* shown in Figure 5.14, adds energy by accelerating the air radially. Impellers are popular on smaller engines because a single impeller replaces several rows of blades in an axial-flow compressor, making them much less expensive to build. There is, of course, a drawback to this less expensive technique. The direction of the air must be turned from heading radially, back to flowing along the axis of the engine. This is the source of a significant energy loss. This loss in efficiency is deemed unacceptable in larger jet engines.

MULTISTAGE COMPRESSORS

A compressor stage, made up of 10 to 12 rotor/stator stages or a single impeller, can only do so much to compress the air. The solution

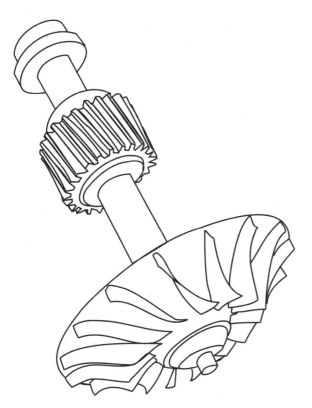

Fig. 5.14. An impeller connected to a shaft.

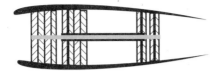

Fig. 5.15. Illustration of a two-shaft, axial-flow compressor.

The General Electric GE-90, which powers the Boeing 777, has a total compressor pressure ratio of 23:1.

to this problem is to add multiple compressors, usually called the *low-pressure* and *high-pressure* compressor sections. In principle, for an axial-flow compressor, you could add more rotor/stator stages. However, as the air compresses and slows down, the rotation speed of the shaft becomes too high. So most commercial jet engines have multiple concentric shafts. A two-shaft engine is illustrated in Figure 5.15. The same thing can be done with centrifugal compressors. Some engines have two impellers, while others have an axial-flow compressor as the low-pressure compressor and the impeller used for the high-pressure compressor.

Burners

One difference between the compression process in a jet engine and an internal combustion engine is that the compression of air is continuous in a jet engine. After the air is compressed, fuel is injected and burned in the *burner,* or combustor. The burner is merely a kind of firebox where the air-fuel mixture is burned. Like the compressor, this is a continuous process.

For best combustion efficiency, the postcombustion temperature is kept as high as possible. Current temperatures at the end of the combustion chamber are on the order of 2800°F (1500°C). This temperature is too hot for typical construction materials, so the burner must be cooled. Bleed air is brought in from the compressor stage and used to form a film covering the inside walls of the burner. The holes for the bleed air are clearly shown in Figure 5.16, which is a photo of a burner removed from an engine. The hot combustion gases never have time to burn through this constantly replenished supply of cool air.

The energy content of fuels is different and the costs vary. For example, electricity and gasoline cost about $0.05 per kilowatthour and peanut butter costs $0.54 per kilowatthour.

It is in the burner that the energy is given to the air to produce propulsion. But before the energetic exhaust can be allowed to escape to produce propulsion there is some work for it to do. Some of its energy must be extracted to power the compressor. This is done by the turbine, which follows the burner.

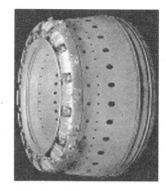

Fig. 5.16. An axial burner. (*Photo courtesy of NASA.*)

Turbines

The turbine looks quite a bit like a single stage of a compressor, only here the first set of blades that follow the burner are fixed and do not rotate. These are called the *turbine vanes*. They are followed by a rotating set of *turbine blades* which drive a shaft connected to the compressor. The arrangement is illustrated in Figure 5.17. The purpose of the turbine vanes is to turn the exhaust into the turbine blades. This allows for greater energy transfer to the turbine blades.

A turbine is the reverse of the compressor. The air expands and cools through each turbine stage, removing energy from the air. The rotating turbine turns the shaft connected to the rotors or impeller in the compressor. There must be as many turbine sections as there are compressor sections. So a jet engine with two compressors, a low- and high-pressure section, will have two turbines, each powering one of the compressors with a separate shaft. The turbine/shaft/compressor combination is referred to as a *spool*. Most large jet engines are *two-spool engines,* meaning that they have a two-stage compressor driven by a two-stage turbine. This was illustrated in Figure 5.15.

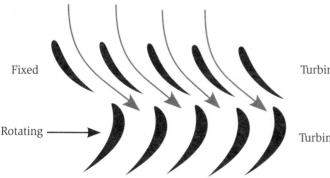

Fixed

Rotating

Turbine vanes

Turbine blades

Fig. 5.17. The turbine with its vanes and blades.

Rolls Royce uses three spools on its larger engines, which increases the efficiency of the engine but also the complexity and cost.

Although the exhaust loses some of its energy, and thus becomes cooler, going through the turbine, it is still very hot. The first vane/blade stage of the turbine sees temperatures similar to those in the burner, on the order of 2800°F (1500°C).

So, this turbine stage requires special cooling, including the film cooling from bleed air as is found in the burner. A turbine blade is shown in Figure 5.18. Notice that it is hollow, which is to allow internal cooling air, and there are small holes on the surface. These holes allow a cool air pocket to form around the surface of the blade. This pocket is thin but allows the blades to survive the hot temperatures.

The pressure change across the turbine goes from high to low pressure. In the compressor the pressure went from low to high. Because of this, unlike the compressor, there is little problem with the turbine blades stalling. The pressure change across a turbine stage, therefore, can be much greater than the pressure change across an axial-flow compressor stage. Even with the energy loss, the gas leaving the turbine still has a great deal of energy and can be used for propulsion.

The Turbojet

The simplest form of the jet engine is the turbojet shown in Figure 5.10. Basically, a turbojet is the turbine engine with a diffuser and a nozzle. The diffuser works to "condition" the air before it enters the compressor. The compressor is optimized for a certain airspeed. Con-

Fig. 5.18. A turbine blade. (*Photo courtesy of NASA.*)

ditioning by the diffuser helps bring the airspeed to the compressor to its optimum speed, regardless of the speed of the airplane. A typical speed entering the compressor might be half the speed of sound (Mach 0.5). So, in a transport that cruises at roughly Mach 0.8, the diffuser slows the air down considerably. When standing still at the end of the runway, the diffuser speeds up the air. This might lead you to believe that the diffuser is an active device. It is the compressor that "demands" how much air must be sucked into the engine. For most jets, the diffuser is thus passive, making sure the air is uniform when it hits the compressor at the right speed. At supersonic speeds it is important that the diffuser slow the incoming air to subsonic speeds as efficiently as possible.

The Sophia J-850 is a very small jet engine. It weighs 3.08 lb (1.4 kg) and produces 18.7 lb (85 N) of thrust. It is a fully functional turbojet engine designed for instruction and model airplanes.

At the other end of the turbojet is the nozzle, which "conditions" the exhaust gas as it exits the engine. The ideal situation is to expand the gas back to atmospheric pressure such that it exits at the greatest possible velocity. This gives the greatest thrust. The design of the nozzle depends on the pressures and velocities of the gas after they leave the turbine.

There are two fundamental problems with the turbojet. First, the turbojet produces thrust with a very high exhaust-gas velocity. We have shown that this requires more power and is thus inefficient. Another problem is that the higher the exhaust-gas velocity the more noise the engine produces. This noise is unacceptable today and FAA noise standards do not permit turbojets to operate at many airports.

Older airplanes, such at the Boeing 707, that used turbojets, are now a rare sight at most airports. The Boeing 727 and 737 originally had turbojets, and many of the earlier versions of these airplanes cannot be flown into many airports because of noise restrictions. In the case of the Boeing 737 the airplane has gone through two major redesigns to improve efficiency and noise. Because of its unique requirements, the Concorde uses turbojets. It is noisy and is considered a "gas-guzzler" by any standard.

The Concorde burns about 500 lb (225 kg) of fuel per passenger seat per hour. The Boeing 777 burns only about 35 lb (16 kg) per passenger seat per hour.

Jet Engine Power and Efficiency

The engine power developed by a turbojet, and thus the amount of air drawn into the engine, is dependent on the amount of fuel injected into the burner and not dependent on the speed of the airplane. If more fuel is added, more force is put on the turbine, causing more air to be brought in by the compressor. It is a characteristic of jet engines that the engine power and the thrust are approximately only dependent on the throttle setting. Thus an engine can develop full engine power and thrust sitting on the runway or at cruise speed.

Now here comes the interesting part. The propulsive power is the thrust times the speed of the airplane. So although the engine may be developing full power and thrust as it starts to roll for takeoff, it is producing almost no propulsive power. This is shown in Figure 5.3. Since the wasted power is the difference between the engine power and the propulsive power, almost all of the power is wasted! We have said that the propulsive efficiency is the propulsive power divided by the engine power. This is almost zero at takeoff and increases with speed. One sometimes hears that jet efficiency increases with speed. This is the source of that increase in efficiency. As will be seen in the chapter on airplane performance, the fact that a jet's propulsive power increases with speed while that of the piston-driven engine is roughly constant affects how these planes make climbs and turns.

> Although a jet engine may be developing full power and thrust as it starts to roll for takeoff, it is producing almost no propulsive power.

Turbojets have another advantage over piston-driven airplanes. Because of the design of the diffuser, the amount of air they take in does not depend strongly on altitude. Therefore, they are able to fly high where the parasitic power is greatly reduced while still developing full power. This enhances the efficiency of jet airplanes.

The Turbofan

Earlier, in our discussion of efficiency, we noted that in order to optimize the efficiency of a jet engine one wants to accelerate a large amount of gas but to as low a velocity as necessary to produce the needed thrust. The nature of turbojets limits the amount of air that

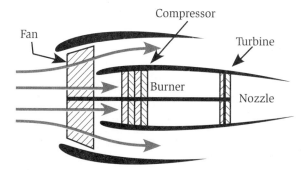

Fig. 5.19. In a turbofan, much of the air bypasses the core.

can be processed. The solution has been the introduction of the *turbofan* engine, as shown schematically in Figure 5.19. The turbofan engine is designed around a turbine engine, but with much more of the energy produced in the burner being converted into mechanical energy by the turbines. Most of this energy is used to turn a large fan in front of the engine. The fan is very much like a propeller, but with 30 to 40 blades instead of just 2 to 4. The large fan accelerates a large amount of air, at a much lower speed than the exhaust of a turbojet producing the same thrust or power. Thus it is much more efficient.

It is important to understand that the air that goes through the fan does not go through the core of the engine but goes around the outside of the core. This can be seen in the photograph of a turbofan in Figure 5.20. Clearly most of the air that goes through the fan bypasses the core. The ratio of the air that goes around the core to that which goes through the core is called the *bypass ratio*. Typical engines today have bypass ratios of about 8:1, meaning that eight times as much air goes around the core as goes through it. Ideally the velocity of the exhaust gas and the air from the fan would be the same. In such a situation, with a bypass ratio of 8, about 90 percent for the thrust would come from the fan and 10 percent from the exhaust of the turbojet powering it.

An additional benefit, and a necessary one, is that the lower exhaust velocity produces less noise. Jet engines today are much quieter than they were 30 years ago. The fanjet also gives engine designers a means for increasing thrust while increasing efficiency. They can increase the mass flow through the engine while decreasing

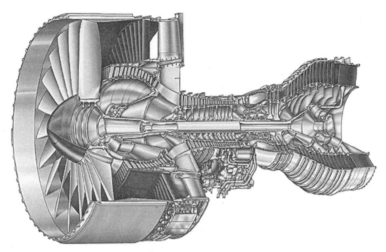

Fig. 5.20. A turbofan engine. (*Photo courtesy of Pratt and Whitney.*)

The engines on the Boeing 777 have a diameter that is within inches of the fuselage diameter of a Boeing 737.

the output velocity. The result of this is that the 100,000-lb thrust engines on the Boeing 777 have a fan so large that the engine's diameter is within inches of the fuselage diameter of a Boeing 737. One can fit six seats and an aisle in one of these engines, though these would be very uncomfortable seats. Figure 5.21 shows a photograph of the engine of a Boeing 777 next to a service truck. This gives a feeling for their size.

Unfortunately, there are practical limits to the size of the engines. But there are always clever engineers out there who find ways to expand these limits. The engines used on the Boeing 777 were unimaginable just two decades ago.

To be certified a jet engine must survive the "bird-strike" test. The engine must survive the impact of a chicken shot at it with a special cannon. The same cannon is used for testing airplane windshields.

Turbofans are the engine of choice for commercial transports and business jets. Military jets also use turbofans but must compromise efficiency if the airplane is to be capable of supersonic flight. The huge inlets of the fanjets are a detriment when it comes to supersonic drag. Military fighters typically use engines with bypass ratios on the order of 3.

The Turboprop

A turboprop follows the same concept as a fanjet. The excess power in the turbine is used to turn a propeller, rather than a fan. Figure

Fig. 5.21. The engine of a Boeing 777.

5.22 shows an example of a corporate turboprop airplane. A propeller cannot be directly attached to this shaft from the turbines because it turns at too high a speed. So a gearbox or *reduction gear* is used to reduce the rotation speed to the propeller. At lower flight speeds the propeller is more efficient than the fanjet. Since it has a larger diameter, it gets its thrust from accelerating more air at a low speed. This we know is the most efficient way to get propulsion. Propellers become less efficient at high speeds because the effects of compressibility, which will be covered in the next chapter, are introduced. Special propellers have been designed to operate at these higher speeds but have not seen use except on experimental aircraft.

Turboprops are used on smaller commuter aircraft and have found a growing market in smaller aircraft. They have also found a growing use in the general-aviation and corporate airplane market. Their big advantage is that they are able to produce greater power than the equivalent piston-driven engine with much less noise and maintenance. Turboprop engines are much lighter than piston engines of the same power but are also more expensive.

Henry Ford tried to be a leader in airplane manufacturing. In 1927, the Ford Tri-motor became the first successful commercial airplane.

Fig. 5.22. A turboprop engine.

Thrust Reversers

When you land in a large jet, you need to decelerate the airplane. As with a car, brakes can be used. But, as we shall see in Chapter 7, "Airplane Performance," the energy that the brakes have to absorb is astounding. Wouldn't it be nice if we could just turn the engines around and have them push the other way to decelerate the airplane? Certainly when you land in a big jet it seems as if they have done just that. A jet engine can redirect its thrust with a *thrust reverser* to accomplish the equivalent result. Though it may seem like it, this does not mean that the engine works backward, that is, blowing gas out the front. What a thrust reverser does is to divert the gas in the jet exhaust and send it forward, i.e., reversing the direction of the thrust. This is illustrated with the *clamshell-type* reverser shown schematically in Figure 5.23. In this figure, the air from the fan is rerouted out through the cowling while the nozzle blocks the core air

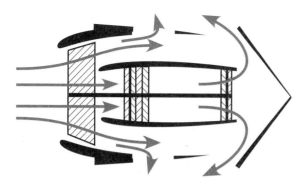

Fig. 5.23. **The thrust reverser partially turns the exhaust forward to produce negative thrust.**

and forces it forward. The net result is that the engine produces some negative thrust to slow the airplane down.

A thrust reverser can only redirect a portion of the air forward. The net forward thrust is just a very small component of what it can produce in the forward direction. But even if the net forward thrust were zero with thrust reversal, the large, non-thrust-producing engines produce a great deal of drag. This alone would help slow down the airplane on landing.

> In 1933, William Boeing wanted to purchase a luxury airplane. He purchased an airplane manufactured by Douglas. In 1940, he upgraded to another Douglas airplane.

Thrust Vectoring

Sometimes jet nozzles can be pointed in a particular direction other than straight back. This is known as *thrust vectoring.* Figure 5.24 illustrates how a hinged nozzle can redirect the jet exhaust from horizontal to an angle. The idea, similar to a thrust reverser, is to redirect the jet to any desired direction. The Harrier is an extreme example of thrust vectoring. The Harrier is able to hover by directing all of the thrust down. The exhaust jet is routed to four nozzles (shown in Figure 5.25) that can swivel from horizontal, for forward thrust, to vertical, for hovering. The Harrier also uses the high-pressure gas from the engine for attitude control as shown. Modern fighter designs, such as the Lockheed-Martin/Boeing F-22 shown in Figure 3.36, use thrust vectoring to increase maneuverability.

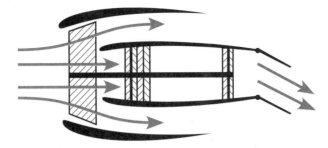

Fig. 5.24. With thrust vectoring the thrust can be redirected to increase maneuverability.

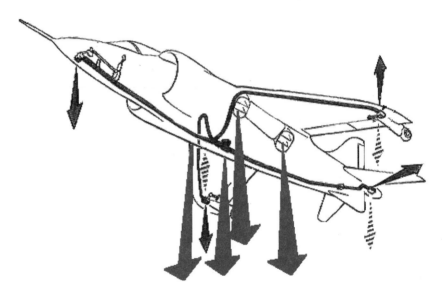

Fig. 5.25. Thrust vectoring of the Harrier in hover.

During the Falkland Islands war between Britain and Argentina, thrust vectoring was found to be extremely effective in combat. Pilots used the thrust vectoring of the Harrier, shown in Figure 5.25, originally designed for vertical or short takeoff and landing, to improve maneuverability. A pilot in straight-and-level flight could quickly direct some of the thrust down and literally make the airplane hop in the air. This extra maneuvering ability allowed the Harrier to avoid air-to-air and surface-to-air missiles at the last minute, by hopping out of the way.

Thrust Augmentation

Military fighters and interceptors sometimes need extra power, called *military power*. One solution is to install a larger engine. But a larger engine weighs more and so is not a practical choice. Instead, some military engines "augment" their thrust. The principle is to add more power to the air after the turbine has removed most of the power to drive the compressor and fan. Fuel injectors are added between the turbine and the nozzle, which injects fuel to mix and burn with the excess oxygen. This device is called an *afterburner*. The energy added by the afterburner is transferred to kinetic energy through the nozzle. The net effect is an increase in thrust and a loss of fuel efficiency. A fighter on afterburners may only be able to maintain this power level for 15 minutes or less.

Visually, an afterburner is impressive. Because the combustion is taking place just before the nozzle, the flame actually extends through the nozzle and out the rear of the engine. You can see the flame when the afterburners are working. As an example, Figure 5.26 shows an SR-71 Blackbird on afterburners. The flame is easily visible. The bright spots are called *diamond shocks*. Because there is so much energy left in the exhaust, afterburners are incredibly noisy.

> The YF-22 (Figure 3.36) is the first airplane that can cruise supersonically without the use of afterburners.

Wrapping It Up

Aircraft propulsion systems involve the same physics as wings. In the case of the propeller, it is nothing more than a rotating wing. The

Fig. 5.26. The SR-71 on afterburners. (*Photo courtesy of NASA.*)

propeller accelerates air back, which pulls the airplane forward. The same is true of a jet engine. The jet accelerates air back to produce thrust. The efficiency of the engine is determined from a compromise between the amount of air flowing through the engine per second and the jet velocity. For subsonic speeds the fanjet maximizes the airflow through the engine and is the most efficient for transports traveling just below the speed of sound. Fanjets are fuel-efficient and relatively quiet. At supersonic speeds the turbojet is more efficient.

Next we look at high-speed flight. There are profound differences in flight at or above the speed of sound where information cannot be communicated forward and the compressibility of air becomes very significant.

CHAPTER 6

High-Speed Flight

Supersonic fighters, commercial transports, bombers, and the Concorde fly at speeds greater than most small airplanes. The speeds at which they fly introduce some additional physics that are not present in low-speed flight. In particular, the compressibility of the air becomes significant. We have said that in the understanding of lift air is to be considered an incompressible fluid because of the low forces involved with low-speed flight. But air is compressible. All fluids, even water, are compressible on some level. When air compresses, its density changes. This is a phenomenon inherently different than in low-speed flight, where the air's density remains essentially unchanged as it passes over a wing, fuselage, or other parts of the airplane.

Figure 6.1 shows a fighter flying at just below the sound speed. The air flowing over and under the wing has expanded, lowering the density and temperature to cause water vapor to condense. In high-speed flight one must understand where, when, and how the density of the air changes.

Mach Number

Low-speed flight is *subsonic* flight. High-speed flight can be broken into three basic categories: *transonic, supersonic,* and *hypersonic.* As

Fig. 6.1. Transonic F-18 with shock wave over wing. (*Photo courtesy of the U.S. Air Force.*)

their names imply, the categories are related to the speed of sound (sonic). As stated before, the Mach number is the airplane's speed in units of the speed of sound. At Mach 1 the plane is going exactly at the speed of sound. Subsonic refers to speeds below Mach 1. Transonic refers to speeds approaching Mach 1. Commercial transports, most military transports, bombers, and business jets fly at transonic speeds. Supersonic refers to speeds above Mach 1, and is usually left to fighters and interceptors for short bursts. Hypersonic refers to speeds of high Mach numbers. At present, the only hypersonic vehicle is the Space Shuttle during reentry.

The North American X-15 (Figure 6.16) reached a speed of 4534 mi/h (7295 km/h) or Mach 6.72.

Mach number also relates to the apparent speed of the air as seen by the airplane. The airplane travels at a single Mach number with respect to the air at a distance. The airflow around the airplane is traveling at different speeds and thus has a different local Mach number. In transonic flight significant parts of the airflow, relative to the aircraft, are both subsonic and supersonic. In the case of supersonic flight virtually all of the air relative to the airplane is supersonic. Finally, in hypersonic flight the vehicle is going so fast that certain additional physical phenomena creep in. This is discussed in a later section.

Why is the compressibility of the air important? In Chapter 2 the idea that air is being diverted down to create lift was emphasized. The lift on the wing was shown to be proportional to the amount of air diverted down times the vertical velocity of that air. The amount of air that is diverted per second was determined by considering the volume diverted. The analogy of a scoop was introduced to demonstrate this. But this assumes that the air density is constant. If density changes, then the amount of air in a given volume changes. This can confuse our intuition of the behavior of air.

A supersonic bomber, being developed by the United States in the 1960s, was lost in a midair collision when it was flying formation for a publicity photo.

Lift Is Still a Reaction Force

High-speed flight, like low-speed flight, requires that air be diverted to create lift. This basic principle does not change with the introduction of compressibility. Ultimately, as an airplane flies overhead, it diverts air down as it passes, regardless of its speed. But there are some changes as to how this occurs at high speeds. Recall that in low-speed flight there is upwash in front of the wing due to circulation. As speed increases, upwash begins to disappear. Upwash was possible because air transfers information at the speed of sound, which is faster than the wing at low speeds. Thus the air is able to transfer information in front of the wing and the air is able to adjust for the oncoming wing. There is upwash and the airflow separates before the arrival of the wing. As the wing moves faster, there is less time for the air ahead to move out of the way of the wing. Once the airplane becomes supersonic, upwash ceases. Also, an almost instantaneous compression wave known as a *shock wave* forms in front of the wing. Shock waves are discussed in detail a little later.

During WWII the United States was producing 5500 aircraft per month.

Understanding shock waves and their drag and power consequences is somewhat complicated. It helps to first understand some of the fundamental properties of supersonic air. During the next few sections a basic primer on supersonic airflow will be presented.

Compressible Air

Because of the small forces involved at low speeds we said that air is to be considered an incompressible fluid. If we were to flow such a fluid through a pipe with a constriction in it, what would we see? We know the answer from the discussion in Chapter 1. As the fluid comes to the constriction, the velocity increases while the static pressure (measured perpendicular to the flow) decreases. So as velocity goes up pressure goes down, and because the fluid is incompressible the density and temperature remain the same. In general, for an incompressible fluid the velocity and pressure change in opposite directions and the density and temperature remain constant.

As the speed of the air increases, the forces become significant and the compressibility of the air becomes significant. Now the density of the air can change and thus its volume no longer remains constant. So let us look at what happens when a compressible fluid such as air flows through a pipe with a constriction in it. At the constriction, or venturi, things look quite similar to the incompressible case, but now the density and temperature can change. As long as the air is traveling at less than Mach 1 at the venturi, the velocity increases and the static pressure decreases, and density and temperature decrease.

If the air is traveling greater than Mach 1 entering the venturi, the velocity decreases and the static pressure, density, and temperature increase. In general, at high speeds where the compressibility of air becomes significant, the velocity of the air changes in the opposite direction to that of the pressure, density, and temperature.

In general, at high speeds where the compressibility of air becomes significant, the velocity of the air changes in the opposite direction to that of the pressure, density, and temperature.

An analogous situation to the flow of compressible air is that of traffic. As the density of cars increases, speeds decrease. At the maximum density we have a traffic jam and a minimum in speed. As the density decreases again and the distance between vehicles increases, so do speeds.

As we discuss later in this chapter and in greater detail when discussing supersonic wind tunnels, there is one big difference in the behavior of supersonic and subsonic air. The difference is that supersonic air cannot communicate upstream (forward in the rest frame of the wing) and can only communicate a very short distance directly above a point on a wing. This will have a profound effect on the behavior of air flowing over a wing at transonic or supersonic speeds.

There is an additional element in supersonic aerodynamics that does not exist in subsonic aerodynamics. This is the formation of shock waves.

Shock Waves

Compression in air can happen over such a small distance that it forms a shock front or *shock wave*. The dimensions on which the air density changes are so small that the change is essentially instantaneous. In supersonic flight a shock wave occurs when air must suddenly change speed and/or direction. Figure 6.2 shows a picture of shock waves on a Space Shuttle model in a supersonic wind tunnel.

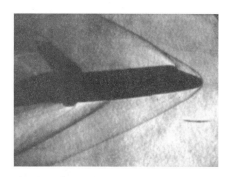

Fig. 6.2. Shock wave on a model of the Space Shuttle.

There are two types of shock waves of interest involved with flight: *normal* (meaning perpendicular) shock waves and *oblique* (at an angle) shock waves. Normal shock waves are perpendicular to the direction of flight and are seen primarily on the surface of transonic wings or in pipes. They are caused by an abrupt change in density and pressure. Figure 6.3 shows what happens across a normal shock wave. Before the shock wave, the air is traveling at greater than Mach 1. Behind a normal shock wave, the air is subsonic and the air's density has increased.

Oblique shock waves are formed at an angle with respect to the oncoming air and occur when supersonic air must be turned. Because a supersonic airplane is traveling so fast, the air has no chance to move out of the way as it does in subsonic speeds. Therefore, the moment the air hits the leading edge of the wing it must turn. The air turns almost instantaneously and forms the oblique shock wave. However, the shock wave forms at a given angle depending on the angle it must be turned. Figure 6.4 illustrates an oblique shock

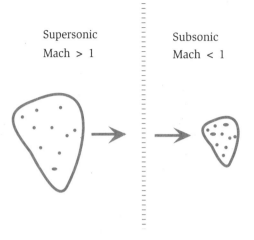

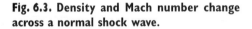

Normal Shock Wave

Fig. 6.3. Density and Mach number change across a normal shock wave.

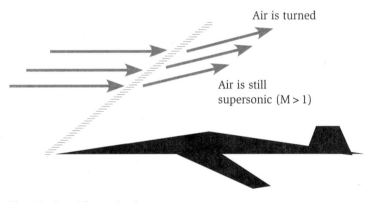

Fig. 6.4. An oblique shock wave.

wave. As with a normal shock wave, the air density increases and the air's velocity decreases across an oblique shock wave. But the changes are not enough for the air to become subsonic, as in a normal shock. Therefore, the air behind an oblique shock wave remains supersonic relative to the aircraft. There are rare exceptions to this rule.

All supersonic objects create shock waves. Normal shock waves cause a higher change in density than oblique shock waves. Therefore, supersonic aircraft are designed to avoid producing normal shock waves, since the greater the change in density across a shock wave the greater the energy loss and the greater the drag. This is accomplished by making the nose and wing leading edge sharp. Blunt noses lead to energy-consuming *bow shocks,* which are a combination of a normal shock wave on the very nose joined to an oblique shock wave a little ways back. Bow shocks are avoided by putting sharp noses on supersonic airplanes.

The SR-71 was designated the RS-71 until President Johnson accidentally reversed the letters in a public announcement. Rather than embarrass the President, the designation changed.

Shock waves in air have known density, pressure, and velocity jumps. For a given shock angle, all of these properties can be found in tables. Thus, for a given airplane geometry the shock angles and pressures are easily determined. Supersonic flight is actually easier to analyze than subsonic flight. In low-speed aerodynamics, engineers must rely on complicated equations to solve for the pressures over a vehicle. In supersonic aerodynamics an engineer can use published tables. However, being able to compute the pressures on the vehicle more easily does not translate into making the problem of supersonic flight easier. Now the penalty of supersonic flight will be discussed: wave drag.

Supersonic flight is actually easier to analyze than subsonic flight.

Wave Drag and Power

As discussed in Chapter 2, the faster an airplane goes the greater the amount of air that is diverted, the smaller the vertical velocity of the downwash, and the smaller the induced power for the same load. Thus the induced power is small for very fast airplanes. The down side is that the parasitic power goes as the speed cubed. Now, there is an additional demand for power at high speeds. The demand is to overcome *wave drag.*

Shock waves travel with the aircraft. Before supersonic flight over land was banned, *sonic booms* (a sound like an explosion heard by those on the ground) were a frequent occurrence around military bases. The shock wave is a persistent phenomenon, which travels along with the aircraft and extends miles from the airplane. This means that the airplane is not only doing work locally to change the airflow but it is affecting the air miles away! This results in an increase in wave drag, and an increase in the required power. The extra power needed to overcome wave drag is one factor that makes supersonic flight so difficult. The design of supersonic aircraft must account for the power required overcoming wave drag, and airplanes that fly supersonically are optimized to reduce it.

Wave drag is more complicated than induced drag and parasitic drag. Unlike the latter two contributions to drag, wave drag is not a simple function of speed but is a complicated function of Mach number. For example, doubling the Mach number, and thus the speed, the power needed to overcome wave drag may increase by less than a factor of 3, whereas the parasitic power would increase by a factor of 8. In this case the wave drag has increased by less than 50 percent. Remember power is drag times speed.

So why does the wave drag decrease so slowly? The drag caused by oblique shock waves depends on the angle the shock wave makes with the direction of the airplane's travel. The more perpendicular the shock wave the greater the drag. As the Mach number increases, the angle of the shock wave decreases. This is illustrated in Figure 6.5. Thus there is a counteracting effect of increasing Mach number. The shock strength increases with Mach number, but because the shock angle is smaller the wave drag does not increase very fast. Since the angle of the shock is a function of the body angle, supersonic aircraft have very sharp noses and leading edges so that the oblique shock wave is produced at as small an angle as possible.

An additional consideration of wave drag is that it is a function of the air density. The greater the air density the higher the wave drag and thus the greater the power required

Can a person in free fall reach a supersonic speed? At 10,000 ft a person would have to be traveling at 750 mi/h (1230 km/h). A person in a dive would have difficulty reaching even 500 mi/h (820 km/h).

Fig. 6.5. As the Mach number increases, the angle of the shock wave decreases.

to overcome it. Thus, aircraft do not fly supersonically at low altitudes.

Transonic Flight

Commercial transports fly in the Mach 0.8 to 0.86 range, just below the speed of sound. This speed is not chosen arbitrarily. It is based on the presence of wave drag. But, if the airplane is flying at a speed less than the speed of sound, how can there be wave drag?

A wing diverts air down. In bending the air down, it creates lower pressure on the upper surface of the wing, which causes the air to accelerate. This topic has been covered in Chapter 2 and will not be repeated here. However, at speeds approaching the speed of sound the air that is accelerated over the top of the wing becomes locally supersonic.

When air flows over the top of a subsonic wing, it accelerates to the point of greatest curvature of the air. At this point the pressure is the lowest and the speed of the air is greatest. From that point to the trailing edge of the wing the airspeed decreases and the pressure increases in order to match the pressure of the air at the trailing edge. This is the *trailing-edge condition.*

The picture is quite different for the air flowing over the top of a transonic wing. The air accelerates as before, but by the time it reaches the point of maximum curvature it is traveling at greater than Mach 1. As the air continues to bend, because it is traveling faster than the communication speed of air, it is not able to effectively pull air down from above. Thus the density is substantially reduced, causing the pressure to continue to go down and the velocity to increase. This situation leaves the wing with the problem of how to meet the trailing-edge condition. The solution is the formation of a normal shock wave as shown in Figure 6.6. At this shock wave the pressure and density increase abruptly and the velocity of the air goes below Mach 1. After the shock wave the air can slow down further and the pressure continue to increase to meet the trailing-edge condition.

Northrop designed a high-speed flying wing, the XP-79, which was a flying ram. Its objective was to slice off the tail of the opponent with its leading edge.

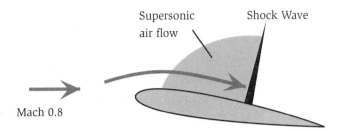

Fig. 6.6. A transonic airfoil accelerates the air to supersonic speeds and then forms a shock wave.

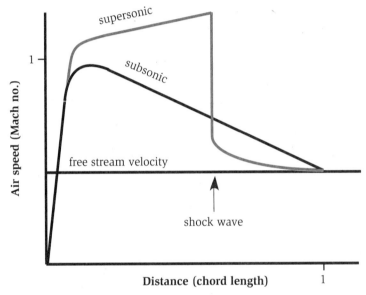

Fig. 6.7. The airspeed over a transonic and a subsonic airfoil.

Figure 6.7 shows the airspeed, in units of Mach number, of the air over a subsonic and a transonic wing. The subsonic airfoil is traveling at a speed just below the *critical Mach number* such that the air never reaches supersonic. The transonic airfoil is just above the critical Mach number so the air becomes supersonic. The Mach number of the subsonic airfoil decreases after the peak while on the transonic airfoil it increases, until the shock wave.

So how does the wing know where to put the normal shock wave? Let us first assume that the normal shock wave forms at the trailing edge in order to meet the trailing-edge condition. What we would find is that the force caused by the pressure difference across the shock wave is higher than the force of the wave drag. Thus, the shock wave will move forward on the wing. As it does, the pressure difference decreases until the wave moves to a place on the wing where the force from the pressure difference just equals the force due to wave drag. If the airplane were now to increase its speed the force due to wave drag would increase and the shock wave would move toward the trailing edge. At some speed the normal shock wave will in fact reach the trailing edge.

Look again at the picture of the fighter flying at transonic speed in Figure 6.1. In the region where the air is supersonic with increasing speed, the pressure, density, and temperature are decreasing. At a point before the normal shock wave the air has cooled enough to cause condensation, producing the cone of fog above and below the wing. The backside of the cone is a flat surface. This is the location of the normal shock wave where the pressure and temperature increase and the condensation disappears. One may ask why there is a normal shock wave on the bottom of the wing. The fighter has almost symmetric wings and since the angle of attack is so small at transonic speeds, there is a reduction in pressure and acceleration of the air on both the top and the bottom of the wing. It is just that the acceleration and reduction in pressure are not as great on the bottom. Commercial jets that fly transonic speeds are designed so that the normal shock wave forms only on the top of the wing.

> The difference in density across the shock wave on the top of a wing can sometimes affect sunlight so that one can actually see the shock wave.

In transonic (and supersonic) flight, the velocity of the air over the wing continues to increase until the normal shock wave is reached. Because of this, the center of lift is farther back on the wing than in subsonic flight. For a typical wing in subsonic flight the center of lift is about one-fourth chord length back from the leading edge of the wing. That means that the wing produces 50 percent of the lift by that point. The moment the wing becomes transonic the center of lift moves farther back. As the speed increases further, the center of lift continues to move back. At very high-speed flight, the center of lift can move as far as to a half chord length back from the leading edge.

In the early attempts at breaking the sound barrier, the jump in the center of lift caused the wing to pitch down. This reduced the acceleration of the air over the wing causing it to become subsonic again. The center of lift would then jump forward on the wing and the wing would pitch up, becoming transonic. This process would repeat itself rapidly until the wings broke off. The X-1 made it through this transition because of a better wing design that made it strong enough to enter supersonic flight.

As the airplane flies faster the shock wave gets stronger. The result is a rapid increase in wave drag, and thus more power is required. The rapid rise in drag, and thus increase in the power required, as one approaches Mach 1 is called the *Mach 1 drag rise*. The design of the wing and body of the airplane is optimized to operate at the "elbow" (or sharp increase) of the power as a function of Mach number, shown in Figure 6.8. Note that the power has a local peak at Mach 1 and then drops off before continuing to rise again. Above Mach 1 an oblique shock wave occurs at the leading edge of the wing. As stated before, as the speed increases above Mach 1 the oblique shock wave bends back at a decreasing angle and thus the increase in *wave drag* power grows more slowly than the increase in parasitic power. Some early supersonic airplanes had to enter a dive to transition from subsonic to supersonic flight. Only after reaching supersonic flight could the airplane level off and maintain supersonic flight with the available power.

Before the oil embargo of 1973, the Boeing 747 was designed to operate above Mach 0.9. However, the fuel costs forced a lower operating speed. One design change in the 747-400 model was to reoptimize the wing to fly at Mach 0.85.

Wing Sweep

In Chapter 3 wing sweep was discussed. In transonic and supersonic flight a swept wing is necessary to reduce wave drag. In 1935, a gathering of top aeronautical scientists from around the world gathered in Rome and showed results of high-speed analysis and wind-tunnel experiments. One German result went largely unnoticed. This result was that a swept wing reduced drag at high speeds.

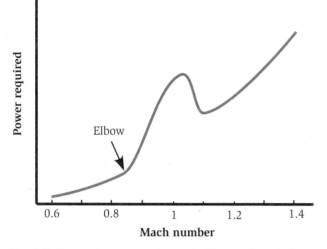

Fig. 6.8. Power increases dramatically as the airplane passes through Mach 1.

The reduced drag of a swept wing results from reducing the *effective* Mach number of the wing. The effective Mach number is the Mach number the wing sees perpendicular to the leading edge, as illustrated in Figure 6.9. A nonswept wing will experience the sharp rise in the power requirement (Figure 6.8) at a lower Mach number than a swept wing. At supersonic speeds wing sweep also helps to reduce the strength of the oblique shock wave from the leading edge of the wing.

The increase in drag at Mach 1 was not understood at the time of the Rome conference. It was later discovered that the presence of the shock wave, shown in Figure 6.6, caused the airflow to separate, increasing form drag. This is illustrated in Figure 6.10. The three airfoils in Figure 6.10 represent three different flight Mach numbers. In airfoil *a,* the flight Mach number is only enough to accelerate the air to supersonic speed on the top of the airfoil. In airfoil *b,* the flight Mach number is

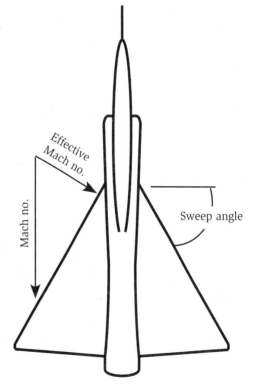

Fig. 6.9. Effective Mach number on a swept wing.

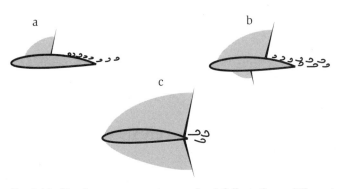

a   b

c

Fig. 6.10. Shock waves on a transonic airfoil at three different Mach numbers.

such that the airflow becomes supersonic on both the top and bottom surfaces, although to a lesser speed on the bottom surface. Airfoil c shows a Mach number near Mach 1 where the shock waves on both top and bottom surfaces have become strong enough to move to the trailing edge and produce a substantial, drag-producing wake. This is the situation of the jet fighter in Figure 6.1. The discovery of the airflow separation led to better airfoil designs for transonic and supersonic flight. The wake due to separation of the air is indicated on all three of the wings in the figure.

For supersonic aircraft, there is a simple relationship relating the angle of sweep to the design supersonic speed of the airplane. This is illustrated in Figure 6.11, which shows the sweep angle as a function of Mach number. The purpose of sweep for supersonic aircraft is to keep the effective Mach number at the wing leading edge at or below Mach 1 (Figure 6.9). It does not take sophisticated military intelligence to determine the supersonic operating conditions of an adversary's airplane. All one has to do is look at the sweep angle. For example, a Mach 2 airplane will have a sweep of 60 degrees.

Area Rule

The amount of wave drag on supersonic aircraft is a function of the size of the aircraft. To illustrate this point, imagine throwing a small

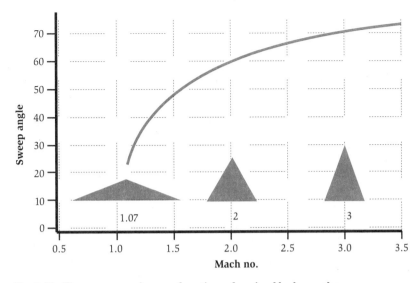

Fig. 6.11. The sweep angle as a function of cruise Mach number.

pebble into a still pond. Small waves propagate away from the entry point of the pebble. Now, repeat the experiment with a big rock. The waves are naturally much larger. The larger waves have more energy than the smaller waves from the pebble. In supersonic flight, the larger the disturbance, the more energy goes into the waves. So supersonic aircraft should be thin and sleek.

A supersonic airplane flying at 60,000 feet can produce a sonic boom that reaches about 30 miles to either side of the flight path.

Some sophisticated analysis performed in the 1940s and 1950s showed that wave drag is proportional to the cross-sectional area (area seen looking at the airplane from the front) of the airplane. At the nose of the airplane the effect of the wave drag grows as the fuselage cross section increases to include the canopy, etc. But when the wing is reached, the cross-sectional area grows dramatically, which causes a large increase in drag and power. The solution is to put a "waist" into the fuselage to maintain a constant cross-sectional area. That is, the area of a slice through the wing and fuselage will have the same area as a slice through just the fuselage either before or after the wing. The maintaining of a constant cross section to reduce wave drag has become known as the *area rule.* Figure 6.12 illustrates the area rule. In the figure fuselages *a* and *d* have the same wave drag because

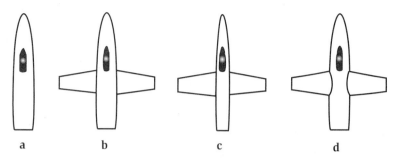

a   b   c   d

Fig. 6.12. What the area rule means to fuselage design.

Fig. 6.13. Two T-38 Talons with fuselage "waist." (*Photo courtesy of the U.S. Air Force.*)

d has a waist to compensate for the wings, giving it the same cross-sectional area as *a*. Fuselage *b* has higher wave drag due to the absence of a waist. Fuselage *c* has the same wave drag as *a* and *d* but at the cost of a small fuselage everywhere.

The area rule is used in the design of most modern fighter aircraft. A notable airplane with an obvious use of the area rule is the T38 Talon, shown in Figure 6.13.

In the discussion above, a simplification was made. To illustrate the area rule we used the cross-sectional area as seen from the front, that is, the area of a slice perpendicular to the axis of the airplane. In reality, the cross-sectional area must be held constant along a slice whose angle is a function of the airplane's design Mach number. Illustrating this complicated result is beyond the scope of this book.

Eight days after the end of WWI the world speed record was set at 163.06 mi/h (262.36 km/h).

Hypersonic Flight

When the Mach number gets high (above about Mach 5), several things happen (Figure 6.14). First, the aerodynamics becomes Mach number independent. This means that for analysis purposes simple assumptions can be made and, in fact, the analysis of idealized, hypersonic flight is the easiest of all aerodynamic analysis. Figure 6.15 shows an artist's conception of a hypersonic airplane. The only known existing hypersonic aircraft is the Space Shuttle during reentry. The X-15, shown in Figure 6.15, explored hypersonic flight in the 1960s, reaching an unofficial speed of Mach 6.7. After the record-making flight, the airplane was retired due to heat damage from the flight. Speculation abounds about the possibility of a supersecret hypersonic spy plane. However, there is only circumstantial evidence to support such a rumor.

The second change that occurs in hypersonic flight is that the energy transfer of the fast vehicle to the surrounding air becomes so great that the air chemistry begins to change. Oxygen and nitrogen

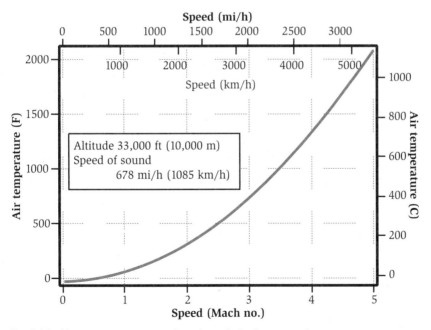

Fig. 6.14. Air temperature as a function of airplane speed.

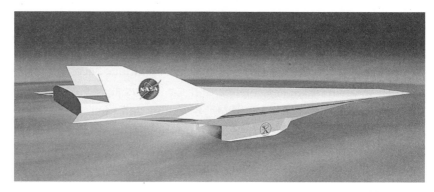

Fig. 6.15. Depiction of Hyper-X hypersonic vehicle. (*Illustration courtesy of NASA.*)

molecules begin absorbing energy by breaking up, or dissociating. This has many implications for the design of hypersonic aircraft. First, the changing air composition can affect the aerodynamics. This was observed in the Space Shuttle, where the direction and attitude stability was predicted to be higher than it actually was.

Pilots of the X-15 flew so high they were given astronaut wings.

Fortunately, enough of a safety margin was designed into the spacecraft so that this shortcoming was not catastrophic. The second major implication is on skin heating, which will be discussed in the next section.

Sir Isaac Newton invented hypersonic analysis using trigonometry. In 1687, Newton was asked to produce better aerodynamic shapes for cannon artillery projectiles. With no previous work in the field, Newton had to invent his own theories. He reasoned that particles of air collide with the surface and then follow the surface after the collision. The result of his analysis is now known as Newton's sine-squared law, which has proved to be a very good predictive rule for hypersonic aerodynamics. However, the first hypersonic flight took almost another three centuries to realize. Of course, Newton did not understand the implications to hypersonic flight. He was trying to solve a much lower speed problem, for which his theory was flawed.

Skin Heating

Thermal protection requirements of hypersonic aircraft are also affected by the dissociation of the air. Vehicles traveling at high Mach numbers will experience extremely hot gases. Some of this is due to temperature increases across shock waves and some is from skin friction. The high-temperature air will burn right through any normal material. The Space Shuttle uses ceramic tiles for thermal protection. The dissociation of air molecules actually helps in keeping the vehicle cooler. It takes energy from heat to break the chemical bonds of the molecules. Thus heat energy is converted to chemical energy and the surface temperatures do not get as hot as would otherwise be predicted. However, they still get very hot, so the surface must be protected.

Extremely high-speed flight is experienced during reentry to the atmosphere. The Space Shuttle, as well as Apollo and Soyuz capsules, must all endure very high heat upon reentry. When the Space Shuttle first hits the atmosphere, it is traveling at approximately 14,000 mi/h (23,000 km/h). The thin air that slams into the nose of the Space Shuttle converts kinetic energy to heat. In theory, the air that impacts the nose of the Space Shuttle will reach over 36,000°F (20,000°C), which is about four times the temperature of the sun! Can this really happen?

When the air reaches high temperatures, it goes through complex changes. Some of the energy of impact goes into breaking chemical bonds rather than creating heat. Oxygen dissociates and ionizes. The impact is so great, in fact, that the ionized gas that develops around a vehicle reentering the atmosphere prevents radio communication with the outside. This is what is known as the *reentry blackout* experienced by all spacecraft since the first successful atmospheric reentry (the Russian Sputnik). Rather than having skin temperatures reaching 36,000°F (20,000°C) the temperature is closer to one-fourth that value, but still sunlike temperatures.

With these high temperatures most things that reenter the atmosphere burn up. "Shooting stars," or meteors, are nothing more than small meteorites, which burn up as they skip

On Oct. 3, 1967, the X-15A-2 (Figure 6.16) was outfitted with an experimental ramjet. The heat was so intense, three of four explosive mounting bolts exploded and the fourth failed, causing the dummy ramjet to separate from the aircraft.

Fig. 6.16. The rocket-propelled X-15. (*Photo courtesy of NASA and the U.S. Air Force.*)

across the Earth's atmosphere. Old satellites usually never get as far as the ground before completely burning. So, how do the Space Shuttle and the command modules reenter at equivalent speeds?

The Apollo command modules had a special carbon-based surface on its base. This surface burned off slowly as the craft reentered the atmosphere. The burning heat shield resulted in two effects. The first was that the burning consumes energy, and thus heat, from the air. The second is that the by-products were swept away, taking heat along with them. The astronauts were thus kept cool behind this heat shield. This form of skin cooling is called *ablation*. The problem with ablation is that it is not reusable.

Special tiles were designed for the Space Shuttle that are extremely poor heat conductors. The tiles absorb heat very slowly. When the surface of the tiles reaches the high reentry temperature, the tiles radiate heat out to maintain a constant surface temperature. But the longer the Shuttle experiences the heat, the deeper the heat will penetrate into the tile. So the tiles must be thick enough to prevent the heat from reaching the aluminum skin before the heat load is removed on landing.

> Hypersonic vehicles do not have sharp noses and wings because there must be enough material to absorb the heat. Sharp objects will burn off at high speeds.

Note that airplanes that fly at slower supersonic speeds experience substantial skin heating. The Concorde fuselage expands as much as 10 inches due to heat from skin friction in cruise. The SR-71 experiences such heat that the top of the wing is corrugated on the ground. The thermal expansion of this surface in flight will remove the corrugation to produce a smooth surface. The SR-71 fuel tanks must also take into account thermal expansion. On the ground, when the SR-71 is cold, the fuel tanks have gaps that leak fuel. By the time the SR-71 has reached cruise, the fuel take has expanded to close all the gaps.

The Concorde fuselage expands as much as 10 inches due to heat from skin friction in cruise.

Wrapping It Up

Newton's laws are still applicable in describing lift on high-speed airplanes. But high-speed flight involves some additional physics such as shock waves, wave drag, and high temperatures. The shock waves arise because the vehicle travels faster than the air can pass information: the speed of sound. High temperatures result from the energy exchange of the fast vehicle with the still air.

Next we look at the overall performance of airplanes and consider tradeoffs that occur when designing for specific missions.

CHAPTER 7

Airplane Performance

Airplane performance has increased spectacularly in the first 100 years of flight. Improvements in wing design, engine performance, and structural design have led to increased range, speed, endurance, and number of passengers carried. Let us pretend we have lived through the last 100 years and traveled regularly from New York to Los Angeles. Figure 7.1 shows how transcontinental air transportation has changed over the years. The trip has gone from over 2 days in length to about 6 hours. The major reductions in time needed to cross the country result from improvements in airplane performance. For example, night flight, which began in 1930, caused a 20 percent reduction in travel time. Before this time, passengers would transfer to trains when it got dark and resume air travel farther along the line in the morning. Though not shown in the figure, there was a slight increase in travel time in 1973 due to the oil embargo and a shift to more efficient cruise speeds. As discussed in the previous chapter, the oil embargo caused wings to be redesigned for slightly slower flight in order to conserve fuel.

Figure 7.2 shows the increase in the number of passengers that could be carried by a passenger airplane through the years. In 1927,

U.S. TRANSCONTINENTAL TRIP TIME
NEW YORK - LOS ANGELES, SCHEDULED AIR SERVICE

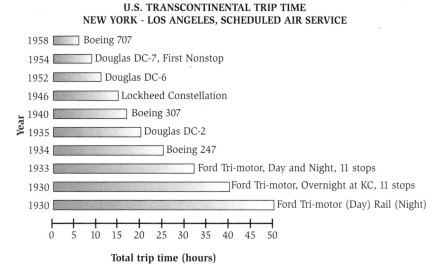

1958 Boeing 707
1954 Douglas DC-7, First Nonstop
1952 Douglas DC-6
1946 Lockheed Constellation
1940 Boeing 307
1935 Douglas DC-2
1934 Boeing 247
1933 Ford Tri-motor, Day and Night, 11 stops
1930 Ford Tri-motor, Overnight at KC, 11 stops
1930 Ford Tri-motor (Day) Rail (Night)

Year

0 5 10 15 20 25 30 35 40 45 50

Total trip time (hours)

Fig. 7.1. Transcontinental travel time from 1930 to 1958.

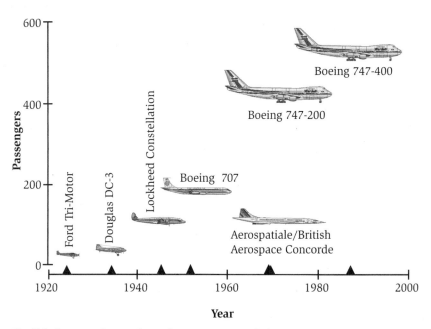

Fig. 7.2. Increase in number of passengers carried.

the Ford Tri-Motor could carry only 11 passengers. The Boeing 747-400 of today can carry almost 50 times that number.

In this chapter you will learn how power, minimum drag, and other factors are used to determine airplane performance. In the sections that follow we discuss the performance of an airplane in powered flight from takeoff to landing. But before that we will prepare you by introducing the lift-to-drag ratio, the glide, and indicated airspeed.

Lift-to-Drag Ratio

There are several parameters that are fundamental to understanding performance. These parameters do not necessarily improve our understanding of how or why airplanes fly but are a useful aid to understand airplane performance. The most important *aerodynamic* parameter is the *lift-to-drag ratio,* often referred to as *L over D* and written L/D. Anyone interested in airplanes has likely heard these words at one time or another. The L/D combines lift and drag into a single number that can be thought of as the airplane's efficiency for flight. Since lift and drag are both forces, L/D has no dimensions, which means that it is just a number with no units. A higher value of L/D means that the airplane is producing lift more efficiently.

A high-performance glider has an L/D of 60:1; an albatross, 20:1; a Boeing 747, 15:1, and a sparrow, 4:1.

In still air, the L/D is the glide ratio, which is discussed in more detail below. You can determine the L/D of a toy balsa-wood glider by measuring its glide ratio, which is the ratio of the launch height to the distance flown (see Figure 7.3). This ratio is the L/D of the glider. It is unlikely that this value of L/D will be the maximum value, but one reflecting how the trim is set for the glider.

There are two ways to look at L/D. If you are an engineer designing an airplane, you have flexibility over both lift and drag. However, for a pilot, in straight-and-level flight the lift equals the weight, so maximum L/D simply means minimum drag. In this book we take the perspective of the pilot and assume the lift is a constant, unless otherwise stated. It is worth looking at L/D from the engineer's perspective first.

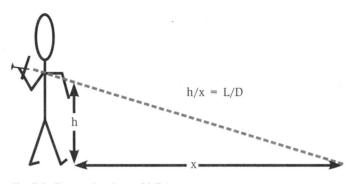

Fig. 7.3. Determination of L/D.

LIFT-TO-DRAG RATIO FROM THE ENGINEER'S PERSPECTIVE

When designing an airplane, the engineer has the ability to change both lift and drag in calculations, in simulations, and in wind-tunnel tests. As an example, when a model is tested in a wind tunnel, the airspeed is kept constant for a set of measurements and the angle of attack is changed. Thus the lift and the drag are varying at the same time. In this situation, unlike in real flight, the maximum L/D is neither at maximum lift nor at minimum drag, but where the ratio is a maximum. These tests are then performed at a variety of airspeeds. With the data gathered this way, the engineer can calculate the performance at different loads and speeds.

The airplane designer must choose a cruising airspeed, and work with various weights and wing parameters such as area and shape. The optimum angle of attack for the wing is then chosen to yield the optimal L/D for cruise conditions. These choices are made to maximize range, endurance, or whatever particular criterion is most important. Flying at an airspeed, or angle of attack, other than this optimum will result in lost performance.

Once the maximum L/D is determined, and the angle of attack at which it occurs, other performance parameters of the airplane begin to fall out. We shall see this in the discussion of ceiling, range, endurance, climb, and turns.

In general, the airplane designer wants the highest maximum L/D possible under the constraints given on wing size, weight, etc. As said before, the higher the L/D the more efficient the airplane. Thus, L/D can be thought of as the aerodynamic efficiency of the airplane.

Glide

A pilot of a powered airplane must be prepared for the loss of power. Contrary to what many think, an airplane does not fall out of the sky if the engine stops. In fact, an airplane can fly quite a distance without power. Without an engine, the airplane is just a poor-performing glider. So, upon loss of power, what should the pilot do?

The pilot may want to maximize the amount of time in the air. This will give more time to search for an emergency landing field, more time to attempt to restart the engine, and more time to communicate with air traffic controllers. To achieve maximum time in the air, or endurance, the objective is to minimize the rate of altitude loss. Since the altitude loss is the source of power keeping the airplane flying, the best endurance will occur at speed for minimum power required (Figure 2.13). At this speed the rate of descent is lowest. On the other hand, the speed is slow and the airplane does not cover much ground as it descends.

What if the pilot is less interested in the time remaining in the air but wants to glide for the largest distance, say, to make it to a better place to land? In this case the pilot wants to increase the glide ratio, which is the distance traveled per loss in altitude. As was introduced with the toy glider example earlier, the distance covered divided by the change in height is the L/D of the airplane. Thus, the maximum glide ratio is equal to the maximum L/D. Since the weight of the airplane is constant, the maximum L/D translates to minimum drag (Figure 2.16). The speed for a glide of maximum range (minimum drag) is typically about 20 percent faster than the speed for greatest endurance (minimum power).

> A commercial jet can glide for about 100 miles (160 km) if it loses its engines at cruising altitude.

> In the case of an engine failure, reducing the weight can increase the airplane's range and endurance. This reduces the load on the wing and thus the induced drag. In WWII flying movies there is sometimes the dramatic scene of the crew throwing excess weight overboard on the return flight due to a shot-up engine.

Gliders have glide ratios of 25 to 60:1. A glide ratio of 25:1 (read "25 to 1") means that for every 1000 ft an airplane descends it travels 25,000 ft horizontally. That is about 5 miles! A typical airliner has a glide ratio on the order of 16:1, while small propeller-driven airplanes have a glide ratio from 10 to 15:1. The glide ratio of the Space Shuttle is only 4:1. It has been said that the Space Shuttle glides like a bathtub.

A pilot is trained to know the speed to be flown when the airplane loses power. Part of transitioning to a new airplane is memorizing the new critical speeds associated with that airplane. But we have seen that power and drag are functions of altitude. Does the pilot need to know critical speeds for every altitude? As you will soon see in the section on Indicated Airspeed, nature has made life a little easier for the very busy pilot.

OUT OF FUEL

On July 23, 1984, in Ontario, Canada, a Boeing 767 ran out of fuel. An error was made converting from the English system of measure to the metric system. The airplane did not have enough fuel to complete the trip from Montreal to Edmonton. The Air Canada pilot, Robert Pearson, was a glider pilot and able to bring the 767 down on an abandoned airfield many miles off the flight path.

Let us assume that the Boeing 767 has a glide ratio of 16:1 and was cruising at 32,000 ft (6 miles) when it ran out of fuel. It would be able to glide almost 100 miles before landing and would have almost 30,000 square miles to land.

Indicated Airspeed

We now come to the concept of *indicated* airspeed. In critical maneuvers such as best climb, longest glide, greatest endurance, etc., the pilot must fly at specific airspeeds. The pilot has committed these airspeeds to memory before flying the airplane. But these speeds change with altitude and air density. So how does a pilot make airspeed corrections for different altitudes and densities? As luck would have it, the *airspeed indicator* on the instrument panel does it for the pilot, and many of the important airspeeds are marked on it.

The airspeed indicator does not measure the true airspeed of the airplane. It is really measuring the difference in pressure produced by the air striking the end of the *Pitot tube* and the static pressure, as discussed in Chapter 1 (see Figure 1.10). A Pitot tube can be seen on a wing of every small airplane. There are several on large jets, mounted near the nose, and they can often be seen as one boards the airplane. The Pitot tube is calibrated so that the indicated airspeed and the true airspeed are the same under *standard* conditions at sea level. As the airplane flies higher, there is less air striking the Pitot tube and thus less total pressure for the same speed. So as the density decreases the indicated airspeed decreases. Since the airspeed indicator is calibrated at sea-level standard conditions, the indicated airspeed is lower than the true airspeed at a higher altitude. The airplane is actually going faster than indicated. For the pilot to determine the true airspeed in flight, the indicated airspeed must be corrected for air density, which is a function of altitude and temperature.

One might think that having to make these calculations in flight is a nuisance and that the pilot would like the airspeed indicator to read the true airspeed. But it is the fact that the indicated airspeed is not the true airspeed that makes the pilot's life easier. All the critical airspeeds of normal flight are indicated airspeeds. Therefore, even though the speed changes with altitude for such important procedures as climb, glide, stall, etc., the indicated airspeeds of these maneuvers remain the same. When a pilot who is used to landing at a sea-level airport makes a landing at a high-altitude airport, the indicated approach speed is the same. But the ground is going by much faster. So, when a maneuver requiring a specific airspeed is performed, the pilot flies the indicated airspeed.

We now have the tools to discuss the performance of the airplane. So we will do just that, starting with takeoff and ending with the landing.

The first around-the-world airplane flight occurred from April 6 to September 28, 1924, and started and ended in Seattle.

All the critical airspeeds of normal flight are indicated airspeeds.

Takeoff Performance

The takeoff of an airplane is a fairly simple thing. The airplane accelerates down the runway until it has reached a speed comfortably above the stall speed. The pilot then pulls back on the controls, or *rotates,* and off the airplane goes. The takeoff speed is typically about 20 percent above the stall speed, but it can be as little as 5 percent for some military aircraft. If you were an airplane designer, what other things would you consider in takeoff performance?

The most obvious figure of merit for takeoff performance is takeoff distance. If you want to design an airplane that can take off from a short dirt field you will have to include certain features. If you have unlimited runway, you might design a different airplane. As a general rule, airplanes that have short takeoff distances will fly at lower cruise speeds. Faster airplanes usually need longer runways. Let us examine why this is so.

Since we are concerned with takeoff distance, it is obvious that one can shorten this distance by increasing the engine power or by reducing the takeoff speed. The problem with increasing the size of the engine is that it adds weight and cost. Also, since the parasitic power goes as the speed of the airplane cubed, the increased power will do little for the cruise performance of the airplane. The takeoff speed is decreased by reducing the stall speed, by reducing the wing loading, and by adding high-lift devices such as slats, slots, and larger flaps. These add cost and weight to the wing and can degrade the cruise performance.

YEAR	NY–LONDON TICKET	COST OF A BOEING 747
1970	$316	$20M
1980	$595	$57M
1990	$462	$123M
2000	$436	$180M

The biggest factor in takeoff performance is the weight of the airplane. The takeoff speed is proportional to the square root of the airplane's weight. A 20 percent increase in weight will cause approximately a

10 percent increase in takeoff speed. But the real killer is that the takeoff distance increases with the weight squared. This is a simple consequence of Newton's second law, which states that the acceleration is the force divided by the mass. So, if the weight increases, for constant thrust, the acceleration decreases and it will take a longer distance to reach takeoff speed. Remember that the takeoff speed has increased to exacerbate the problem. So, for example, a 20 percent increase in weight increases the takeoff distance by about 44 percent for a high-powered airplane. But an increase in weight of 20 percent will increase the takeoff distance of low-powered general-aviation airplanes by about 60 percent because of the lower acceleration.

The takeoff distance is also impacted by the wind. A headwind that is 15 percent of the takeoff speed will shorten the takeoff distance by about 30 percent while a tailwind of the same speed will lengthen the takeoff distance by 33 percent. For a small airplane with a takeoff speed of 70 mi/h (112 km/h) this is only a 10 mi/h (16 km/h) wind. This is why airplanes always take off into the wind. Aircraft carriers turn into the wind to launch and recover aircraft.

During the early years of aviation, WWI and earlier, airfields were large, square, or circular fields. The airplanes could not tolerate crosswinds as they do today so they would point into the wind for takeoff. When runways were developed, this meant that airplanes could no longer take off in any direction, depending on the local wind. However, careful airport design will place the main runway into the prevailing winds, and sometimes a secondary runway is built perpendicular to it. With a secondary runway it guarantees that the airplane will at most see a 45-degree crosswind.

Altitude also contributes to takeoff performance. Recall that it is the indicated airspeed that dictates airplane performance. So the takeoff ground speed increases with altitude, though not the indicated airspeed. Here the difference between a jet and a piston-powered airplane is apparent. The thrust, and thus the acceleration, of a jet engine is less affected by altitude. At an altitude of 6000 ft (1800 m) the takeoff distance of a jet is increased by 20 percent over the sea-level distance. The altitude affects the acceleration of a non-turbo-charged, piston-powered airplane, and thus its takeoff distance is increased by about 40 percent at 6000 ft.

Small single-engine airplanes, like the Cessna 172, can take off in distances less than 1000 ft at sea level. The Cessna 172 has a wing loading of only 13.5 lb/ft^2 (66 kg/m^2). The light wing loading contributes to the short takeoff distance. Its available power is moderate, and, as we saw in Chapter 5, the thrust available decreases with speed, and so is a maximum during the takeoff run.

A Boeing 777 has a high wing loading of about 100 lb/ft^2 (490 kg/m^2). This is over seven times that of the Cessna 172. Its ground-roll takeoff distance is on the order of 6000 ft (2 km). In order to comply with FAA regulations for air transports, which require that the airplane be able to fly safely with one engine inoperable, the thrust available for the Boeing 777 must be double what is necessary for a sustained climb. So the thrust available compared to the weight of a Boeing 777 may be double that of the Cessna 172. The thrust available compared to the weight is called the *thrust-to-weight ratio.*

A Boeing 777 cannot take off from a 1500-ft (500-m) grass runway. On the other hand, the Cessna 172 cannot cruise at 500 mi/h (800 km/h). Its top speed is only about 125 knots (225 km/h).

Climb

Up to this point, besides glides, only straight-and-level flight has been discussed. You have probably thought about what happens in a climb to increase altitude. A simplistic answer is that you need to generate more lift, the logic being that as one increases the angle of attack, the lift goes up, and the airplane climbs. When the pilot first pulls back on the controls to start a climb, this is what happens, but only for a few seconds while the airplane is slowing down. The airplane must slow down because no power was added to create this lift. Then, at this lower speed there will be less lift then there was initially. In a sustained climb the lift of the wing is actually less than the weight of the

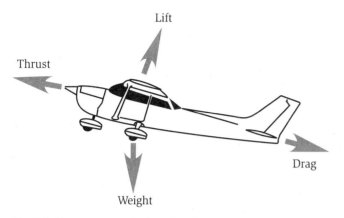

Fig. 7.4. **Forces on an airplane in climb.**

airplane. The forces on the airplane are rotated, except for the weight, as shown in the diagram in Figure 7.4. In this rotated configuration, part of the weight of the airplane is supported directly by the engine. As we will see, the airplane is climbing on the *excess thrust* and *excess power* of the engine.

In a sustained climb the lift of the wing is actually less than the weight of the airplane.

To understand this, let us begin by looking at two extreme situations. First, take the case of straight-and-level flight. The angle of climb is zero and the lift on the wing is the weight of the airplane. Now consider a very powerful jet fighter that can go straight up in a climb. In this case the angle of climb is 90 degrees and the lift on the wing is zero. The engine now supports the weight of the airplane. As the fighter goes slowly from straight-and-level flight to a vertical climb, the load on the wing smoothly changes from the weight of the airplane to zero. During this transition the *lift* produced by the engine goes smoothly from zero to the weight of the airplane.

Now let us consider what happens when a low-powered airplane goes into a climb. Consider a small single-engine propeller airplane flying straight-and-level at full power. In this case the wings produce the lift and the power is just equal to the induced and parasitic power requirements. Now the pilot pulls back a little on the controls and the airplane starts to climb. Part of the engine's power now goes directly into lifting the airplane. This leaves less power to overcome drag and to produce the remaining lift with the wings. Thus the airplane slows down. As the pilot continues to pull back on the controls, the speed is

On February 21, 1979, at Kitty Hawk, former astronaut Neil Armstrong climbed to 50,000 ft in a business jet and set five world records.

further reduced until the *backside of the power curve* is reached. Eventually the wing would stall if the angle of attack were further increased.

There are two climb scenarios of interest to pilots. The first is the fastest climb or the *best rate of climb*. Airplanes fly more efficiently at higher altitudes, so pilots generally want to climb to their desired cruise altitude as quickly as possible. This is the rate of climb most useful to a pilot. The second scenario is the steepest climb, or the *best angle of climb*. Suppose you are in a mountain valley and wish to clear the mountaintops. You would want to gain as much altitude in the shortest distance possible, and thus fly at the steepest angle. These two scenarios lead to different climb paths and airspeeds, which are shown schematically in Figure 7.5.

The best rate of climb for an airplane occurs when the excess power is the greatest. The excess power is the difference between the propulsive power (power providing thrust) and the power required for flight. This is shown in Figure 7.6 for an airplane fitted with either a jet engine or a propeller. The power required is just the power curve discussed in Chapter 2. The arrow connecting the power curve with the propulsive power for the propeller is the excess power. When it is the longest possible, it is at the speed for the best rate of climb. The best rate of climb for a propeller-driven aircraft is at a speed that is very near the minimum drag. One would guess that it would be near the minimum power required, but the variation of the propulsive power with speed causes the best rate of climb to be at a higher speed.

As shown in the figure, the propulsive power of the jet increases with speed. Thus the arrow marking the greatest excess power for the

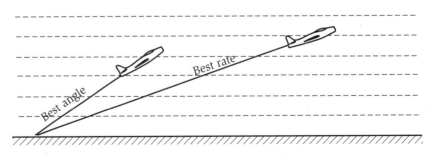

Fig. 7.5. Best angle of climb and best rate of climb.

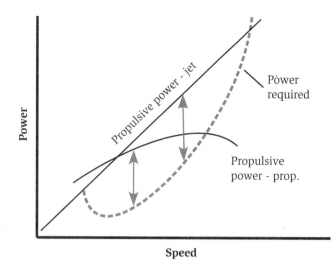

Fig. 7.6. Excess power as a function of speed.

jet is at a considerably higher speed than for the propeller-driven airplane. This is why propeller-driven airplanes slow down to climb while jets maintain their speed or even accelerate a little to climb.

The best angle of climb is achieved with the maximum excess thrust. The reason for this is not particularly difficult but beyond the scope of this book. Take a look at Figure 7.7, which shows the relationship of thrust and drag (required thrust) to the speed of jet and propeller-driven airplanes.

The available thrust of a propeller-driven airplane decreases with increasing speed. Thus, as shown by the arrow, the maximum excess thrust does not occur at the minimum drag but at a lower speed. In fact the best angle of climb occurs just above the stall speed of a propeller-driven airplane. Such an airplane, taking off from a short runway, with an obstacle such as power lines, will clear them by climbing at about the takeoff speed.

As can be seen in Figure 7.7, the thrust of a jet is approximately constant with speed. Thus the best angle of climb for a jet-powered airplane is achieved at the minimum drag.

For a Cessna 172 at sea level the best rate of climb is achieved at an indicated airspeed of 84 mi/h (134 km/h).

Sometimes technology does not rule. During the Vietnam War, 91 percent of all U.S. fighters shot down by antiaircraft fire were aimed at by hand.

The best rate of climb is associated with excess power and the best angle of climb is associated with excess thrust.

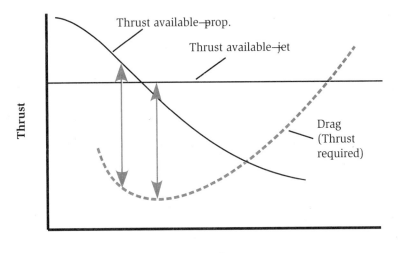

Speed

Fig. 7.7. Excess thrust as a function of speed.

Airplanes need a lot of fuel. A car is about 5 percent fuel by weight; a city bus, 2 percent; a passenger train, 1.1 percent; a freight train, 0.4 percent. A Boeing 747 is 42 percent fuel by weight.

This produces a climb angle of 6 degrees and a rate of climb of 770 ft/min.

The thing to remember is that the best rate of climb is associated with excess power and the best angle of climb is associated with excess thrust. Next we take a look at how high an airplane can climb.

Ceiling

As an airplane climbs, the air becomes less dense. The power available from the engine decreases at the same time that more power is needed to produce lift. What happens is illustrated by Figure 7.8. As the airplane climbs, the minimum speed that the airplane can fly increases. This is because the air becomes less dense so the minimum speed at which the wing will divert enough air without stalling increases. The speeds for the best rate of climb and the best angle of climb are increasing because the minimums in the power and drag curves are shifting to higher speeds. As the figure shows, at some altitude, the minimum flight speed and the two climb speeds meet. At this point, the airplane cannot fly higher and full power and thrust are needed to sustain straight-and-level flight. This altitude is the *absolute ceiling*.

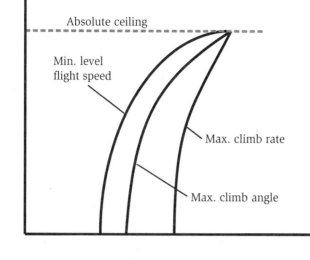

Absolute ceiling

Min. level
flight speed

Altitude

Max. climb rate

Max. climb angle

Speed

Fig. 7.8. The minimum flight speed and climb speeds meet at the absolute ceiling.

As the airplane climbs, its rate of climb decreases. The military defines the *combat ceiling* as the altitude where the best rate of climb drops to 500 ft/min (150 m/min). For some small general-aviation airplanes that is not much above the runway at sea level. The *service ceiling* of an airplane is defined by the FAA as the altitude at which the airplane's best rate of climb drops to 100 ft/min (30 m/min). This is a useful measure of the performance of an airplane. If one flies out of a high airport such as the 10,000-ft (3000-m) airport at Leadville, Colorado, an airplane with a 20,000-ft (6000-m) service ceiling is certainly much more desirable than one with a service ceiling of only 14,000 ft (4200 m). The latter airplane will take much more runway to take off and will climb much more slowly.

Let us look at two high-altitude aircraft that have taken two different approaches for the same mission. The first is the U-2 high-altitude reconnaissance airplane, shown in Figure 3.5. The airplane is an updated 1950s design with a service ceiling over 70,000 ft (21,000 m). This airplane has a high aspect ratio wing to increase lift efficiency by minimizing drag and maximizing L/D. Its engines are not powerful

The world altitude record for an airplane was set on August 22, 1963, at 354,200 ft (107,960 m) by an X-15.

enough to achieve supersonic flight, so in this case the high service ceiling is due to the high L/D rather than a high power-to-weight ratio. The fact that this airplane was slow led to the incident with Gary Powers, when he was shot down over the Soviet Union in 1960. This political catastrophe helped push the development of a replacement reconnaissance aircraft, the SR-71.

Since the U-2 could not fly fast, it was susceptible to antiaircraft missiles. The SR-71 (shown in Figure 3.6) was developed to prevent such possibilities. The SR-71 had to fly at high altitudes and at very high speed. Its published service ceiling is 80,000 ft (24,000 m) and its maximum speed is Mach 3.2 or 2300 mi/h (3700 km/h). In order to be fast, the SR-71 has a high thrust-to-weight ratio. The compromise for high-speed flight is that the L/D is low. So the SR-71 and the U-2, designed for the same mission, made two drastically different choices in their designs.

> The air pressure at 63,000 ft (19,000 m) is so low that water will boil at body temperature.

Fuel Consumption

When we think of fuel consumption for a car, we think in terms of miles per gallon (or liters per 100 km). These are natural units, since cars have odometers and we measure the amount of fuel when we fill up. On the other hand, these units are not appropriate for an airplane. The airplane is flying in a moving fluid. A small airplane in a strong head wind at a low power setting can actually fly backward with respect to the ground, while measuring a substantial positive airspeed.

Pilots are more concerned with how much fuel is on board and how long they can remain airborne. Recall that in Chapter 2 you learned that induced power was proportional to load squared. The pilot of a commercial airplane wants to fill up with as little fuel as necessary. By the end of the flight, the fuel tanks should contain only the FAA required reserves, which should be enough to reach an alternate destination, if necessary. The important parameters in determining the necessary fuel are the anticipated ground speed, which gives the time in the air, and the rate that fuel is consumed. The rate of fuel consumption is measured in units of gal/h (l/h) for small airplanes and lb/h (kg/h) for large airplanes. Unlike a car, the rate of fuel consumption of an airplane depends only on the power setting and not the ground speed. If the airplane will be flying into a head wind, the ground speed will be lower, the time in flight will be longer, and thus more fuel must be carried.

> In 1911, "Cal" Rogers made the first transcontinental flight of the United States, taking 49 days to fly from New York to Los Angeles.

Now, suppose you were an engineer and were going to design an airplane from scratch. You have a variety of engines at your disposal. It should be intuitive that the lower the power output of the engine, the lower the rate of fuel consumption. So you could just choose a small engine, right? Consider a Boeing 777 and a Cessna 172. Can those use the same engine? Clearly the engineer must consider some other criteria for efficiency. Engineers use a parameter called the *specific fuel consumption*. This is simply the rate of fuel consumption divided by the thrust or power produced, depending on the type of engine. The specific fuel consumption is thus a measure of the engine's efficiency. The lower the value, the more efficient the engine. So, when choosing an engine, the engineer must not only consider its weight, the takeoff distance, cruise speed, and ceiling; the specific fuel consumption must also be factored in.

Maximum Endurance

If an engine is burning a certain amount of fuel per hour, how long can it stay in the air? Is there a speed at which the airplane can remain in the air longest? Some airplanes are designed for endurance. For example, airplanes built for surveillance may want to loiter over a particular location for a long time. There is interest in using autonomous aircraft that will relay local communications, such as cellular phones, rather than using expensive satellites. These aircraft are mostly concerned about the length of time they can stay in the air. An example is NASA's Pathfinder airplane, shown in Figure 7.9, which was designed to loiter at extreme altitudes and gather atmospheric data. The aircraft is solar-powered so that the fuel consumption is zero and the endurance in principle is limitless. At night it runs on batteries charged during the day. Let us see how one would determine how to get maximum endurance out of a fuel-powered airplane.

> In April 1949, the Sunkist Lady stayed aloft for 6 weeks and 1 minute (1008 h, 1 min). The airplane was refueled by flying low over a Jeep that passed up fuel cans.

Fig. 7.9. Pathfinder, a maximum endurance, solar-powered airplane. (*Photo courtesy of NASA.*)

In maximizing the time aloft, the speed is not the concern. What is important is the rate at which fuel is burned. The maximum endurance in the air for any fuel-carrying airplane is just at the speed of minimum fuel consumption. For a piston/propeller airplane, the propulsive power is almost directly proportional to the engine power. The engine power, and thus the fuel consumption, is just proportional to the required power for flight. So the speed at which the pilot should fly a propeller-driven airplane for maximum endurance is at the minimum of the power curve. ·

> The speed at which the pilot should fly a propeller-driven airplane for maximum endurance is at the minimum of the power curve.

Things are different for jet-powered airplanes. As discussed in Chapter 5, "Airplane Propulsion," the fuel consumption of the engine is dependent on engine power, not propulsive power. The engine's propulsive efficiency increases with speed and the propulsive power increases with speed, while the thrust of the engine remains constant. This means that the airplane gets more propulsive power for the same fuel flow as the speed increases. Or put another way, for a given propulsive power the fuel flow can be reduced as the airplane speeds up. This relation to speed means that the minimum fuel flow for a jet is not at minimum required power, as it is for a propeller-driven airplane, but at minimum required thrust, i.e., minimum drag. So a jet pilot should fly at the minimum drag speed for maximum endurance. Since minimum drag is a higher speed than minimum power, an airplane with a jet engine will cover more ground flying at maximum endurance than a propeller-driven airplane.

Maximum endurance means maximum time in the air. This is useful for the very few that have a need to stay in the air for a long time. But a more interesting cruise condition is maximum range.

The speed at which the pilot should fly a jet airplane for maximum endurance is at the minimum in the drag curve.

Maximum Range

What is the speed for the maximum range on a tank of fuel? The best range for an airplane is different from the maximum endurance. At maximum endurance the airplane is traveling quite slowly. Now we want to cover as much ground as possible on the available fuel. As with endurance, the conditions for maximum range are different for a propeller-driven airplane and jet.

The world record for a nonstop, nonrefueled flight is 24,987 mi (40,204 km), set by the specially designed Voyager in 1986 while flying around the world.

For a propeller-driven airplane, the propulsive power is roughly independent of speed. For the maximum range one does not want to find the minimum in power, as for endurance, but the minimum in power divided by speed. Power is equivalent to fuel consumption and speed is equivalent to distance traveled. So power divided by speed is equivalent to gallons per mile. But power divided by speed is just drag. So, for the maximum range a propeller-driven airplane flies at the speed for minimum drag. This speed is higher than the maximum endurance speed.

For the maximum range a propeller-driven airplane flies at the speed for minimum drag.

Now let's consider a jet. Recall that a jet flew at the minimum drag speed for maximum endurance. This was because there was a speed term associated with the efficiency of the engine. So the argument goes as before, and we get the result that for maximum range a jet-powered airplane flies at the speed for the minimum drag divided by speed.

The maximum endurance and range for a propeller-powered airplane are at the speeds for minimum power and drag. For a jet the maximum endurance and range are at the speeds for minimum drag (power/speed) and minimum drag/speed.

For the jet we get the result that for maximum range a jet-powered airplane flies at the speed for the minimum drag divided by speed.

We know what speed to fly for maximum range. But the load of the airplane decreases with time. So how does an

Fig. 7.10. Boeing 747. (*Used with the permission of the Boeing Management Company.*)

airplane adjust for this reduction in weight to minimize the fuel consumption on a flight? For small changes in weight an airplane adjusts its angle of attack. But for a large jet, such as the Boeing 747 shown in Figure 7.10, the most efficient way to penetrate the air at a high speed is with the fuselage exactly aligned with the direction of flight. So the pilot wants to maintain the most efficient angle of attack throughout the flight.

To keep the angle of attack constant throughout the flight, commercial jets prefer to adjust for larger weight corrections by reducing the amount of air the wing diverts. This is accomplished simply by climbing to a higher altitude where the air is thinner. So, as the fuel is consumed, the airplane wants to climb to where the air is less dense.

It is straightforward to understand the change in altitude with change in weight. If a large jet were to become 20 percent lighter due to the consumption of fuel, the lift must be reduced by that amount. Since it is undesirable to change the angle of attack of the wings or the speed, the vertical velocity of the downwash is a constant. Therefore, the amount of air diverted must be reduced by 20 percent. This means a 20 percent reduction in air density. So, as a large jet burns fuel, it wants to fly at an ever-increasing altitude so that its weight divided by the air density remains constant. This is called a *cruise climb*.

As an example of the change in altitude required, a large jet initially flying at 30,000 ft (9100 m) would have to fly at about

A KLM DC-2, which was making a commercial passenger flight, came in second in the 11,333-mile race from England to Australia in 1934.

35,000 ft (10,600 m) in order to correct for a 20 percent reduction in weight. Ideally this change in altitude would be accomplished by a slow climb over the entire flight. Of course, for safety reasons, the FAA would not allow this. Unfortunately, for fuel consumption, jets are only allowed to increase their altitudes in 4000-ft (1200-m) steps as they fly.

A pilot can request an altitude change during the flight and thus approach a cruise-climb situation with *a stepped-altitude climb*. Most long flights will perform a stepped-altitude climb, and it is not uncommon for the pilot to announce such a change in flight. The purpose is to decrease the amount of fuel consumed.

Pilots of small airplanes fly at a constant altitude and speed. If you are a pilot of smaller aircraft, this is the easiest condition to maintain. As fuel is consumed and the airplane gets lighter, the angle of attack will have to be adjusted. Unfortunately, changing the angle of attack also changes the airplane's location on the drag curve. The airplane is no longer flying at maximum efficiency. The lost range from the lost efficiency is on the order of 10 percent over that of a long cruise-climb flight. This means that if an airplane could travel 1000 miles (1600 km) in cruise-climb, it would only go 900 miles (1440 km) at fixed altitude and airspeed.

One might ask whether large airplanes fly at the best fuel consumption. The answer is "almost." Take a look at Figure 7.11, which shows the distance traveled per amount of fuel as a function of the airplane's speed. Large airplanes try to fly on the high side of the maximum. They fly at 99 percent of the maximum range since the 1 percent loss in range gives them a several percent increase in cruise speed. This is considered a good exchange.

As early as 1946, the U.S. Air Force was considering nuclear-powered airplanes. In 1955, the NB-36H actually flew a nuclear reactor, although it was conventionally powered.

You should recognize that the range a particular pilot gets from an airplane is dependent on the pilot's particular flying habits and how fast he or she wants to get to the destination. Most pilots of general-aviation airplanes fly considerably faster than the speed for maximum range. This is because the cost of the fuel is a smaller fraction of the total operating cost and a good part of the expense of flying is tallied by the hour.

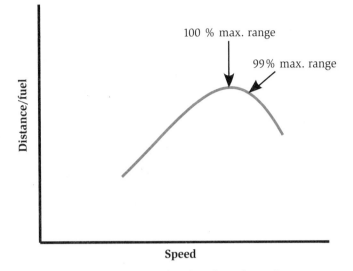

Fig. 7.11. Fuel consumption as a function of speed.

Turns

Now that we have the airplane in flight, let us make a turn. Unfortunately, for commercial and general-aviation airplanes, turns are hardly interesting. If they were, passengers would complain. (However, to someone not familiar with a 2g turn, this might still qualify as interesting.) High-performance turns are primarily the domain of fighter and aerobatic aircraft.

In the discussion of the climb we saw that the engine, not the wing, is lifting the airplane. But it is the wing that makes a turn. Recall that in Chapter 1 we saw that the load on the wing (and the pilot) increases when an airplane goes into a bank. In a turn the *load factor* and stall angle of attack become two critical components in understanding turn performance (refer to Figures 1.6 and 1.7).

An average pilot can withstand about 6 g's for a few seconds without blacking out.

The load factor is just like real weight as far as the wing is concerned. We know that the induced power and induced drag vary as the load squared. So when an airplane makes a 2g turn the induced power and induced drag (which are important in most turns) have gone up by a factor of 4.

With this reviewed, let us look at an easy (not high-performance) turn with a bank angle of 45 degrees. This turn will put a force on the wings and the passengers only 40 percent larger than in straight-and-level flight. That is, the load factor is 1.4. The induced power and induced drag will be increased to about twice their values in straight-and-level flight. The pilot has two choices to compensate for the increased induced power, increase the power available or decrease the parasitic power to compensate. The former implies adding more engine power, the latter reducing the speed and increasing the angle of attack. For most general-aviation pilots a turn is entered at constant power, the pilot adjusts the angle of attack with the elevator, and the airplane loses speed.

In Table 7.1 we look at the performance of an airplane making a 180-degree turn, in a 45-degree bank, at three different speeds. For a given bank the radius of the turn increases as the speed squared while the time to make the turn increases as the speed. For all three speeds in the table the forces felt by the passengers will be the same, though the turns will be quite different. This is shown in Figure 7.12, which illustrates turns at three speeds for an airplane in a 45-degree bank for 20 seconds. The effect of speed on turn performance is quite dramatic.

> The SR-71 takes 8 minutes to complete a 180-degree turn at cruise.

Table 7.1 Turn Performance for 45 Degree Bank at Three Different Speeds

Speed mi/h (km/h)	140 (224)	280 (450)	560 (900)
Radius of turn, mile (km)	0.25 (388)	1 (1.6)	3.9 (6.2)
Time to make 180 degrees	20	40	80

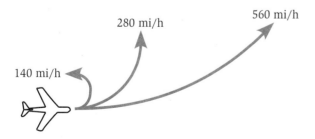

Fig. 7.12. Turns with a 45-degree bank for 20 seconds for three speeds.

If the bank in the example above were increased to 60 degrees, the load factor would be increased to 2, and the induced power and induced drag would be increased to 4 times the values in straight-and-level flight. But the radius of turn and the time to make the 180-degree turn would be reduced by 42 percent. Suppose we increased the bank angle even more. Eventually there have to be some limits, since at 90 degrees the load on the wing and the power required would become infinite.

Suppose you were flying up a canyon (considered a very bad idea) and you wanted to make a tight turn to get out. What is required for a high-performance turn? How does one make a turn of minimum radius?

The minimum-turn radius is limited by three characteristics of an airplane. These are (1) the stall speed with the flaps up, (2) the structural strength, and (3) the propulsive power that is available. Let us look at each of these limits individually.

Stall Speed Limit

What is the stall speed limit? This can be illustrated by considering the extreme of an airplane flying at just above the stall speed (i.e., just below the stall angle of attack). If this airplane tried to make a turn, it could not increase its angle of attack to accommodate the higher load on the wing or it would stall. So this airplane would be unable to turn and thus have an infinite turning radius. If the airplane flew a little faster, it could make a very gentle turn, with a very large turning radius. The very early airplanes were so underpowered that they could not fly much faster than the stall speed, and so flew in this predicament. They could only make slow turns of large radius. The first flights of many, including the Wright brothers, could only be made in a straight line.

Engine power grew rapidly, once flight became a reality. By 1910, the French had built an engine that could deliver 177 hp (179.5 kW).

Following the logic of the previous paragraph, lets see what happens when the pilot is in a turn at a speed twice the straight-and-level stall speed. From Chapter 2 we know that at double the speed the airplane can hold four times the load before it stalls, due to the increase of diverted air and the increase in the vertical velocity of that air. In a bank, this means that the airplane can make a 4g turn (a load factor of 4), which occurs in a 75.5-degree bank. What this shows is that, for any given speed, the maximum

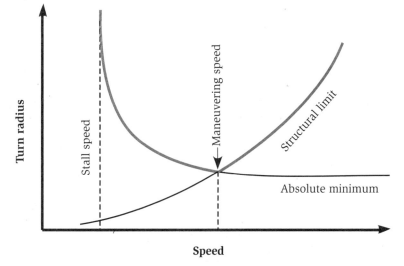

Fig. 7.13. Turn radius as a function of speed.

performance turn is made at a bank angle, and thus wing loading, just slightly less than what would cause a stall.

For a given airplane, the faster one goes the higher the wing loading that can be achieved before stalling in a turn and thus the tighter the turn. This is illustrated by the curve marked *absolute minimum* in Figure 7.13. The slower the stall speed, the more slowly the airplane will be going to achieve the same load factor in a turn and the tighter the turn. So the slower the stall speed the tighter the high-performance turn.

Structural Strength Limit

The second characteristic that limits the turn performance is the structural strength of the airplane. This sets the limit to the load on the wing and is marked *structural limit* in Figure 7.13. For airplanes in the "normal" category, which applies to most general-aviation airplanes, the load limit is 3.8. We just saw that a maximum performance turn, at double the straight-and-level stall speed, exceeds this limit. Thus, the pilot must either reduce the bank angle at this speed or slow down.

If we combine the stall speed limit and the maximum load factor limit, there is a specific condition where the airplane is at both limits. The speed at which this occurs is the *maneuvering speed*. At this speed the wing will stall just before it exceeds the rated structural limits. So in the normal category the tightest turn possible would be made at 1.95 (the square root of 3.8) times the stall speed of the airplane and at a bank angle of about 73.5 degrees. It would be an uncomfortable turn with such a high g force.

> The maneuvering speed of an airplane is the maximum speed for an airplane to use during a maneuver or in turbulent air. As discussed above, at this speed the wing will stall at just the maximum rated load factor, and thus not exceed it. Since the maximum load factor at stall is proportional to the speed squared, the maneuvering speed of an airplane is just the stall speed times the square root of the maximum load factor.
>
> We know that the stall speed of an airplane increases with load. Therefore, so does the maneuvering speed. Thus a more heavily loaded airplane can safely fly faster in a storm than if it were lightly loaded. It is important to note that the maneuvering speed is an indicated airspeed. Thus the true airspeed at the maneuvering speed increases with altitude.

Propulsive Power Limit

Now let us look at the importance of the propulsive power that is available. As the airplane goes into a steep bank, the load on the wing increases and the induced power and induced drag increase as the load squared. So an airplane in a 4g turn experiences an increase in induced power of 16 times. Thus the available propulsive power may not be sufficient for the theoretical tightest turn. When this condition is reached, the maximum performance turn is no longer at the maneuvering speed.

The radar return on the B-2 Bomber and F-117 Stealth Fighter are roughly the same as an eagle.

Power available for both jet and propeller / piston engines decreases with altitude and the induced power for straight-and-level flight increases. There is also an increase in stall speed with altitude. At some point, an airplane that is trying to achieve a maximum performance turn becomes power-limited. Rather than turn at the maneuvering speed, the minimum-radius turn will occur at a somewhat lower speed. This will result in a load factor less than the maximum load factor. So, at higher altitudes, the best turn performance decreases.

There is one caveat that allows tighter turns despite the loss of power available. The pilot can choose to "buy" power with altitude. In other words, in a descending turn, the pilot can supplement the engine power with the power it used to climb to altitude in the first place (like a car going down a hill it first had to climb). This is why fighter airplanes engage at high altitudes but the fight progresses to lower altitudes. The airplanes are using altitude to tighten their turns.

During WWII, the "grasshoppers," airplanes such as the Piper Cub, had the lowest combat losses. In one case, a Cub pilot, jumped by a German fighter, managed to turn, land in a field, and hide the airplane in the time it took the German fighter to turn around.

STANDARD-RATE TURNS

The turning rate is usually not a critical design issue for aircraft other than fighters and specialized acrobatic airplanes. However, all aircraft must be able to perform a standard-rate turn. A standard rate turn for light airplanes is defined as a 3 degrees/s turn, which completes a 360-degree turn in 2 minutes. This is known as a 2-minute turn. For heavy airplanes a standard-rate turn is a 4-minute turn. Instruments, either the turn and slip indicator or the turn coordinator, have the standard-rate turn clearly marked. Light aircraft are equipped with 2-minute turn indicators while heavy aircraft are equipped with 4-minute turn indicators. This is very useful to pilots who are out of visual contact with the ground and for air-traffic control when appropriate separation of aircraft is desired. The pilot banks the airplane such that the turn and slip indicator points to the standard-rate turn mark and then uses a watch to time the turn. The pilot can pull out at any desired direction depending on the length of time in the turn. Note that the standard-rate turn should be well below the maximum turning rate.

Landing

What goes up must come down. So another performance parameter is landing distance. Landing distance is easier to understand than takeoff distance. When the airplane approaches its touchdown, it has a certain amount of kinetic energy $(1/2\ mv^2)$. When it comes to a stop, it will have zero kinetic energy. So the landing distance will be proportional to the touchdown velocity squared. Once again, landing performance benefits from a low stall speed.

Typically, the landing distance of an airplane is shorter than the takeoff distance. This is because the airplane can decelerate with its brakes faster than it can accelerate with its engines. As for most cars, one can stop in a shorter distance than it takes to accelerate to the same speed. There have been many stories of pilots who have landed in a short field but have been unable to take off.

Once on the ground the airplane's minimum stopping distance will depend primarily on its ability to brake. The braking power is proportional to the weight supported by the wheels. On a hard dry surface, the decelerating force from the brakes can be as high as 80 percent of the weight on the wheels. Of course, this value is greatly reduced for a slippery surface. So, for a maximum performance stop the lift must be removed from the wings as quickly as possible to put the weight on the wheels. Thus, as soon as the airplane touches down, the flaps are raised. Modern jets also employ *spoilers* on the top of the wings, which remove part of the lift.

During its return to landing, the Space Shuttle, which has no power during its descent, is a glider with a glide ratio of 4:1.

One key factor in determining the minimum stopping distance is the ability of the brakes to absorb energy. The brakes of a 500,000-lb (227,000-kg) airplane landing at 170 mi/h (270 km/h) must dissipate 30,000 hp (22 million watts) for $^1/_2$ minute or so! The heat from the brakes can be so great that the wheels will literally melt. This energy that must be dissipated makes the use of thrust reversal important to reduce the demands on the brakes.

Commercial airplane tires are filled with pure nitrogen to remove oxygen that can contribute to a fire, in case of maximum braking. Also, the heat will cause the tires to expand and potentially blow up. Some airplanes use special bolts that will separate under high temperatures so that the tire pressure is reduced before they can explode. So normally commercial airplanes do not use maximum braking. But even after normal landings airplanes are required to wait for a certain length of time to let the brakes cool. Flights cannot shorten the time before takeoff without artificially cooling the brakes. Cooling times for normal landings are on the order of 30 minutes. Airlines must make sure to schedule airplane turn-around to accommodate requirement.

Light aircraft do not have a problem with the amount of energy that the brakes must dissipate because their landing speed and weight, and thus their kinetic energy, are so low. In addition, these airplanes are usually landed in a high-drag configuration, such as with full flaps, which will naturally slow the airplane down without brakes. Pilots are always happy when they reach a nice rolling speed by the first runway turnoff without using brakes.

Like takeoffs, wind and altitude affect landings also. In fact, they affect landing performance in exactly the same way as they do takeoffs. Earlier we gave an example of a head wind that was 15 percent of the takeoff speed, shortening the takeoff distance by 30 percent. A head wind that is 15 percent of the landing speed will shorten the landing by the same 30 percent. Likewise, the landing distance at an altitude of 6000 ft (1800 m) will be 20 percent longer than at sea level.

Wrapping It Up

Now that we have discussed performance characteristics we can look at the performance of some different aircraft. Let us consider a small single-engine airplane, a large transport, and a military fighter. The Cessna 172, the Boeing 777, and the Lockheed-Martin F-22 stealth fighter discussed in Chapter 3 will be used as examples here.

The Cessna 172 (Figure 3.10) has very low wing loading as well as moderate L/D, power-to-weight ratio, and specific fuel consumption (fuel consumption divided by the thrust or power produced). With the low wing loading, the Cessna has a very short takeoff distance when compared to the other two examples. However, its cruise speed and range do not compare. This is in part due to the fact that the Cessna 172 carries less than 10 percent of its weight in fuel. Because of the low wing loading, the Cessna's turning radius is much smaller than that of the other airplanes. However, this is not to imply that the Cessna can outmaneuver the F-22, since the rate of turn is low.

The Boeing 777 (Figure 3.35) has a high wing loading and, moderate L/D and thrust-to-weight ratio. It also has a low specific fuel consumption. Because of the high wing loading, the Boeing 777 needs long runways found only at the larger commercial airports. Although it has a maximum L/D similar to the Cessna it has a higher thrust-to-weight ratio. Therefore, the Boeing 777 has a service ceiling about 2 1/2 times that of the Cessna 172. One design goal of the Boeing 777 was to have a long enough range to service trans-Pacific routes. Therefore, efficient engines are used and it carries up to 40 percent of its weight in fuel. Neither tight nor fast turns are necessary or possible with this airplane.

The F-22 (Figure 3.36) has a high wing loading, low L/D (highdrag), high thrust-to-weight ratio, and moderate specific fuel consumption. The high thrust-to-weight ratio allows for extremely steep climbs and high maneuverability. It also gives a high service ceiling, despite the low L/D. Its service ceiling is just a little higher than that of the Boeing 777. The range of the F-22 is low because of the high drag and the desire to trade fuel weight for payload weight. The F-22, like all fighters, will require frequent refueling for long trips.

The next chapter introduces you to the uses of flight testing and wind-tunnel testing. These types of tests are used to determine some of the parameters discussed in this chapter.

CHAPTER 8

Aerodynamic Testing

You have been introduced to many concepts of flight, from lift to performance. One might ask how the designers know that their calculations are correct and, once the airplane is constructed, how the performance is determined. This is the subject of aerodynamic testing, first in a wind tunnel and finally in the air.

Wind-Tunnel Testing

Wind-tunnel tests were performed well before the first airplanes flew. Most of the early wind tunnels were built to perform experiments of fundamental fluid motion. The scientists who built these apparatuses were not interested in flight, but instead the physics of fluids. The Wright brothers can be credited as being one of the first to use a wind tunnel to test aerodynamic shapes. Their ingenuity and accomplishments in testing are still a source of amazement after 100 years. Many pages could be and have been written on these accomplishments, so we will not go into detail here. The important thing to note is that the Wright brothers recognized the need for useful aerodynamic data and so built a wind tunnel and a device that could measure lift.

> The first Boeing 747 completed more than 15,000 hours of wind-tunnel testing.

Before delving into the purposes of wind tunnels, let us explore some of the basic concepts behind the wind tunnel. First, we discuss the subsonic venturi, which is often mistakenly used in descriptions of flight.

Subsonic Wind Tunnels

A subsonic wind tunnel works like a venturi shown in Figure 8.1. A venturi tube is the best example of Bernoulli's theorem, which relates speed and pressure in a tube and when no energy is added to the fluid (see Appendix). As the air reaches the restriction in the tube, the velocity increases. As discussed in Chapter 1, the increase in velocity causes a reduction in the static pressure, measured perpendicular to the direction of flow. Since the forces are low, the air density and temperature remain essentially constant. As we will see later, this is not true in the case of transonic venturis.

A WING IS NOT A HALF VENTURI

You might find it amusing to note that ground school courses often introduce the venturi as an example of how wings fly. The presentation includes moving one of the walls of the venturi so far away as to not influence the other wall (see Figure 8.2). What is left is a wall with a hump. The instructors tell the students that because of Bernoulli's principle this "half venturi" has lift. But you now know that this is wrong. The wall blocks the downwash, so there can be no lift. After leaving the hump, the air is traveling at the same speed and in the same direction as before the hump. As we know, if there is no net change made to the airflow there cannot be lift. So, what do you do when you see this on the FAA written exam? Well, if you want to pass, you will have to give them the answer they want to hear, even though it is wrong!

The simplest of all low-speed wind tunnels is a venturi tube. If you have a household fan and cardboard, you can easily build a small wind tunnel, such as shown in Figure 8.3. The fan is placed so that it draws air through the wind tunnel. The model is placed in the venturi. The cross section of the wind tunnel does not have to be round. In fact, most wind tunnels have rectangular cross sections.

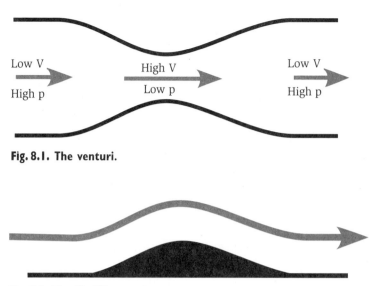

Fig. 8.1. The venturi.

Low V
High p

High V
Low p

Low V
High p

Fig. 8.2. The "half" venturi.

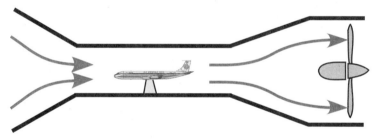

Fig. 8.3. Wind tunnel.

The change in cross-sectional area of the tunnel from the fan section to the test section is called the *contraction ratio.* If you reduce the cross section of your cardboard wind tunnel to one-fifth, a contraction ratio of 5, the airspeed in the test section will be five times the airspeed produced by the fan. So a fan with a 2-ft (61-cm) diameter moving air at a speed of 15 mi/h (24 km/h) would produce 70 mi/h (112 km/h) in a test section about 11 inches (27 cm) in diameter. On the other hand, if you want a test section that is 5 ft in diameter, you will need a fan 11 ft (3.3 m) in diameter. A more practical solution is to increase the fan's airspeed and reduce the contraction ratio to achieve a larger test section at the same speed.

A wind-tunnel model for a major development program can cost millions of dollars.

There are practical problems in building a venturi wind tunnel. You cannot just contract and expand the tube arbitrarily. This is because the air must constrict smoothly to reduce the effects of the walls. To contract too quickly will make the walls act as a block to the airflow, as illustrated in Figure 8.4. The pressure buildup limits the effectiveness of the fan. To expand too quickly after the venturi also causes problem. The air will not be able to follow the walls, causing the flow to separate from the walls. This causes a buildup of pressure, which also reduces the effectiveness of the tunnel. Figure 8.5 illustrates how a venturi wind tunnel should work.

> A rule of thumb for wind tunnels is that the walls should not slope more than 7 percent after the test section.

The wind tunnel just described, where the air passes once though the test section and is then lost, is called an *open-circuit* wind tunnel. A practical problem with such wind tunnels is that all the energy put into air is lost and cannot be recycled. This makes the open-circuit wind tunnels inefficient. Therefore, you generally do not see large venturi-type tunnels. There is one exception, the 80 × 120-ft wind tunnel at NASA Ames Research Center, discussed a little later.

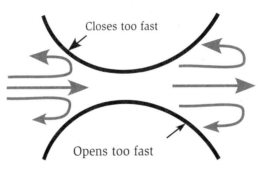

Closes too fast

Opens too fast

Fig. 8.4. A venturi that contracts and expands too quickly.

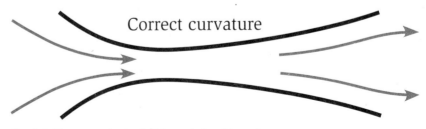

Correct curvature

Fig. 8.5. How a venturi wind tunnel should work.

MEASURING THE AIRSPEED IN THE TEST SECTION

The venturi is an application of the Bernoulli principle. As the air enters the contraction, it speeds up, and the static pressure drops. Since Bernoulli's theorem states that static pressure and airspeed are related, this relationship can be used to determine the airspeed in the test section. Without going through the mathematics, the velocity in the test section is just proportional to the square root of the difference between the static pressure in the test section and the static pressure in the region with the fan. The proportionality just depends on the density of the air and the difference in cross-sectional areas of the two sections. A manometer is used to measure the pressure difference. In its simplest form, a manometer, shown in Figure 8.6, is just a bent glass tube filled with liquid connecting the two regions. The difference in pressure is proportional to the difference in height of the liquid as shown. The wind-tunnel operator only has to monitor the difference in height of the liquid and the air density to know the airspeed in the test section. Kerosene is often used as the liquid because it will not evaporate and is safer to handle than mercury.

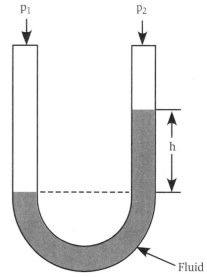

Fig. 8.6. The manometer.

Closed-Circuit Tunnels

The lost power in an open-circuit wind tunnel can easily be recovered if you close the airflow circuit, as in Figure 8.7. The closed-circuit wind tunnel is the most common design for larger tunnels. Once the air is accelerated to operating conditions, the fan only needs to add the power that is lost due to drag on the model and friction on the walls of the wind tunnel.

Be aware that this power loss may be quite significant. Consider a production wind tunnel running many hours a day. The friction and drag on the model can result in considerable heat input to the air and the walls of the wind tunnel. Some

In 1709, Brazilian Jesuit Bartolmeu Lourenco de Gusmao demonstrated the first hot-air balloon to the King of Portugal. Unfortunately, the fire used in the balloon ignited the royal draperies, causing considerable damage.

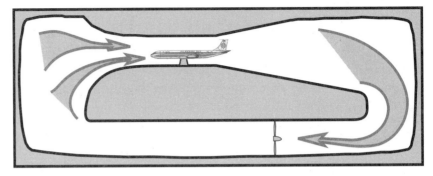

Fig. 8.7. The closed-circuit wind tunnel.

sophisticated wind tunnels have cooling vanes to take this heat out and try to maintain a constant temperature. Other wind tunnels just suffer in the heat. Some tunnels can get so hot that changes to the wind-tunnel models must be performed with insulated gloves.

Transonic wind tunnels are designed to test aircraft at roughly Mach 0.8 to 0.85. These tunnels require much more power than their low-speed counterparts. Since the power loss due to friction goes as the airspeed cubed, much more heat is generated and must be removed.

Most wind tunnels have a single return, as shown in Figure 8.7. There was a time when dual-return wind tunnels were popular. The Kirsten wind tunnel at the University of Washington, Seattle is a dual-return wind tunnel. The layout of this wind tunnel is shown in top view in Figure 8.8. The advantage of such an arrangement is that two smaller motors can be used rather than one large motor. There is also an advantage of size with a dual-return wind tunnel. For reasons beyond the scope of interest of this book, a dual-return wind tunnel can support a larger test section while having a smaller footprint. Smaller motors and a smaller footprint result in a lower cost of construction. The disadvantage is that the two channels of air must meet and become uniform by the time they reach the test section. This causes additional technical difficulties. Figure 8.9 shows a model in the test section of the Kirsten wind tunnel. The test section of this wind tunnel has a cross section 8 ft high by 12 ft wide (2.4 m × 3.6 m).

The NASA Ames 40 × 80 wind tunnel is the largest close-circuit wind tunnel in the United States. In the jargon of wind tunnels a 40 × 80

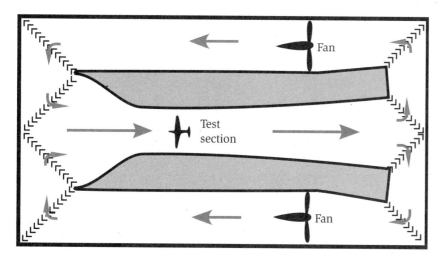

Fig. 8.8. University of Washington Kirsten wind tunnel.

Fig. 8.9. Model in the Kirsten wind tunnel.

wind tunnel has a test section that is 40 ft (12 m) high and 80 ft (24 m) wide. This wind tunnel can produce winds from 0 to 350 mi/h (650 km/h). Six fans and six motors, shown in Figure 8.10, drive the wind tunnel. The railing in front of the upper row of fans gives one the feeling of the scale. You might also be able to make out the three people

Fig. 8.10. Fans of the NASA Ames 40 × 80 and 80 × 120 wind tunnels. (*Photo courtesy of NASA.*)

In 1909, the Wright brothers filed suit against Glenn Curtiss for patent infringement. The Wrights contended that Curtiss' use of ailerons was the same as wing warping. The lawsuit held up development of aviation in the United States until WWI.

standing in front of fan number three. The fans are 40 ft (12 m) in diameter with 15 variable-pitch blades. Each motor is rated at 12 megawatts (18,000 hp).

NASA Ames also has an 80 × 120 (24 × 36 m) open-circuit wind tunnel. The truly impressive test section of this wind tunnel is shown in Figure 8.11. Here winds from 0 to 100 mi/h (160 km/h) can be produced. Both the 40 × 80 and the 80 × 120 use the same fans shown in Figure 8.10. The tunnel can be reconfigured with turning vanes to be either an open- or closed-circuit tunnel. The original tunnel was the 40 × 80, but in the early 1980s a plan was put forth to create the 80 × 120. This major upgrade allows for large wind-tunnel models to be tested in the 40 × 80 and full-scale aircraft to be tested in the 80 × 120.

One thing to note about both open-circuit and closed-circuit wind tunnels is that the test section is usually kept at atmospheric static pressure. Significant pressures can build up in closed-circuit tunnels,

Fig. 8.11. Test section of the NASA Ames 80 × 120 open-circuit wind tunnel. *(Photo courtesy of NASA.)*

particularly in supersonic wind tunnels. Keeping the test section at ambient pressure allows the use of windows for view and photographing the model. Exceptions to this are highly specialized tunnels that have pressurized test sections to increase air density.

The Wright brothers were not the first to use a wind tunnel. However, they were the first to use it for the purpose of understanding flight.

Wind-Tunnel Data

What sort of data is collected in a wind tunnel and how is it used? The most obvious purpose of a wind tunnel is to measure forces and torques, which are twisting forces, on the airplane. Wind tunnels can also be used to measure pressures and airflow patterns on parts of a model. Much of this testing is highly specialized and complicated and so it will not be discussed here. However, there are some interesting things to learn about wind-tunnel testing and how it relates to the concepts introduced in this book.

FORCES

First, we focus on lift, drag, and torques. A typical wind tunnel uses a *force and moment balance* to measure these aerodynamic forces. Figure 8.12 gives an illustration of such a balance. Though the figure shows only one gauge, in reality a force balance makes measurements of all of the forces and torques on the model. In fact there can sometimes be more than one balance used, as can be seen in Figure 8.11.

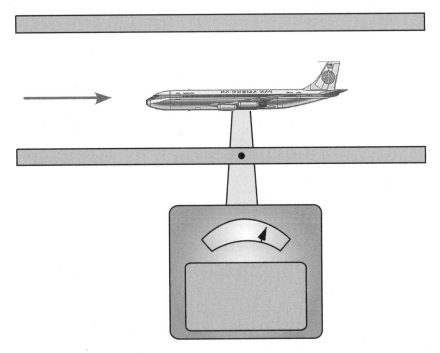

Fig. 8.12. Wind-tunnel force balance.

One question is "are they accurate?" Several factors enter into the accuracy of wind-tunnel measurements. The most important factor affecting the accuracy of measurements is the effect of wind-tunnel walls.

The wind tunnel introduces an artificial constraint, namely, walls. The walls have two effects. The first is that they interfere with the amount of air that can be pulled from above the model's wings and blocks the downwash on the bottom. This latter effect is just like ground effect and is called *wall effects.*

The cost of military aircraft has increased consistently since the first purchase of a Wright Model A. At this rate, the entire defense budget of the United States will buy one airplane in 2050.

Another problem in a low-speed wind tunnel is that air will speed up in a constriction and the airplane model acts like a constriction in the venturi. In other words, the air accelerates as it moves around the model due to the blockage from the model. For obvious reasons, this is called the *blockage effect.* Wall and blockage effects are illustrated in Figure 8.13.

Years of theoretical work have resulted in methods to correct for wall and blockage effects. Unfortunately, wind-tunnel corrections are not completely reliable, so wind-tunnel results must be backed up with flight test. Normal wind-tunnel corrections amount to only a few percent. However, the few percent can be extremely important when trying to predict performance of the final airplane. It is not uncommon for an airplane manufacturer to use the same wind tunnel for all tests because the engineers acquire experience in estimating how a particular wind-tunnel result will relate to actual flight data.

One of the most important pieces of data collected is called the *drag polar.* This is a plot of lift vs. drag, as shown in Figure 8.14. At first glance these data do not look very interesting, but look again. What

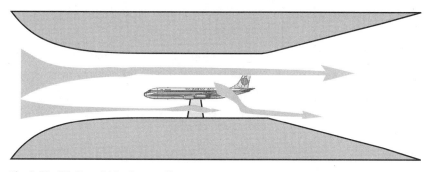

Fig. 8.13. Wall and blockage effect.

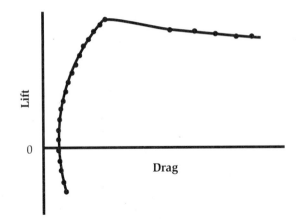

Fig. 8.14. Test data of lift vs. drag.

makes a drag polar interesting is that it can be used to determine the parasite and induced drag as well as the maximum lift before the stall occurs. The minimum-drag value on the graph is just the parasitic drag of the model. The drag measured at any other value of lift is the induced drag plus this parasitic drag. Since the induced drag and lift are available from these data, the wing efficiency can be determined. Notice that at some point the lift decreases and the drag increases dramatically. This of course is the point where the wing stalls and the form drag increases. From this the poststall characteristics of the wing are determined. A great deal can be learned about an airplane from this single plot.

> Artist and inventor Leonardo da Vinci wrote 35,000 words and made 500 sketches on flight.

PRESSURES

A more sophisticated wind-tunnel test will also involve the measurement of pressures on a wind-tunnel model. Tiny holes, called *pressure taps,* are drilled into the model. These taps are connected with tubing to a *pressure transducer,* which is a small electronic device that converts pressure to a voltage to be read by a computer. Some models have over one thousand pressure taps. So the pressure measurement system must be able to scan all of these pressure taps in a short time.

The pressures can be used to determine many things. They can determine flow separation on a surface, as well as being used to calculate local forces. These pressure measurements can also supply validation data to numerical simulations.

FLOW VISUALIZATION

A third type of data collected in the wind tunnel is visualization data. Unfortunately, in a normal wind-tunnel test there is very little to see. You cannot see the wind flowing. Since wind-tunnel test sections are closed to prevent data corruption, you cannot even feel the wind. All you hear is lots of noise from the powerful fans. The only way to see what is happening to the air is to have something visible to blow around.

The Boeing 747 was originally designed to compete against Lockheed's C-5 transport. Boeing lost the military contract but went into production of the airplane as a commercial jet instead.

The most common visualization tool is smoke. The problem with smoke is that in order to see details the wind speed must be very slow. The very low speed can alter the airflow enough to make the smoke results misleading. Another problem with smoke is that in a closed-circuit wind tunnel the smoke builds up after a while.

Another method is to use a mixture of clay with a fluid that evaporates quickly. The clay is very fine and has the consistency of talcum powder. This is painted on the model and then very quickly the wind is turned on and the model positioned. Once the fluid evaporates, the clay is left in a pattern of the surface flow, as shown in Figure 8.15. Areas of flow separation are easy to spot as well as patterns of the general flow around the airplane. In the figure one can see that the last quarter of the wing is stalled. This method is very effective but cannot be used if there are pressure taps, since the clay will clog the holes.

Tufts of yarn are also useful. Very small tufts that minimize the impact on the model can be glued to the model's surface. The tufts follow the direction of the airflow as seen in Figure 8.16 and can be photographed for later analysis.

A fast-growing technique is the use of pressure-sensitive paints. This is a paint that actually changes intensity and shades of a color depending on the local pressure. So it is becoming possible to actually see pressures on the surface of a model. This technology is still under development but may become common in the future. There are many

Fig. 8.15. China clay flow visualization.

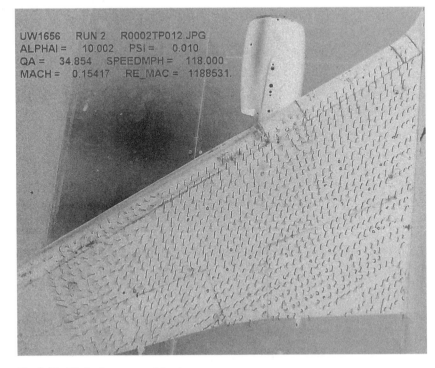

Fig. 8.16. Minitufts on model wing.

other methods of visualization, but to discuss them in detail is beyond the scope of this book.

TAIL ON AND TAIL OFF

If you visited a wind-tunnel test, there is a good chance you would see an incomplete airplane. Probably the horizontal stabilizer will not be installed. A normal wind-tunnel test will include many runs without the horizontal stabilizer. There are two reasons for doing this. The first is to be able to provide an estimate of the errors due to wall effects. The horizontal stabilizer has its own upwash and downwash, which will enter into the measurements. By removing the horizontal stabilizer, the corrections for the wing alone can be determined.

The second reason for removing the horizontal stabilizer is so the effect of the stabilizer on torques on the airplane in flight can be directly measured. As the name implies, a stabilizer is used to stabilize the airplane. The effect of the horizontal stabilizer on the total torque on the airplane determines how stable the airplane will be, how difficult it will be to fly, and where the payload can be positioned, as was discussed in Chapter 4. In analyzing flight characteristics there are many parameters that cannot be determined theoretically. So the wind-tunnel data gives data that can be used to fill in where theory cannot provide a clean answer.

> The sweep on the Boeing 727 wing is 32.5 degrees. United Airlines liked 30 degrees, but Boeing's previous airplanes had 35 degrees of sweep. So the engineers settled on 32.5 degrees.

Supersonic Venturis

In order to understand transonic and supersonic wind tunnels, it is necessary to understand the transonic and supersonic venturi. Here the compressibility of air cannot be ignored, and both the air density and temperature change significantly. Let us first look at the flow of air that is just below Mach 1 in a tube that decreases in diameter with distance. As the diameter of the tube decreases, the velocity of the air increases and the static pressure decreases. Since the forces are now quite high, as the air compresses both the density and temperature of the air increase. Because of this compressibility, the velocity and static pressure do not change as much as they would if the air were incompressible.

> During the Gulf War, U.S. airplane loss rates were roughly the same as during normal training.

If the tube decreases in size enough, the velocity of the air will reach Mach 1. In a constricting tube, the air does not want to go faster than Mach 1. In fact, in a restricting tube, the speed of the air will always change in the direction of Mach 1. If the air before the restriction is going faster than Mach 1, the dynamic pressure and density will increase at the restriction, slowing the air down until Mach 1 is reached. Once Mach 1 is reached, by either accelerating subsonic air or decelerating supersonic air, the pressure will build up, moving the Mach 1 region forward into a smaller radius in the tube.

> In a restricting tube the speed of the air will always change in the direction of Mach 1.

The net result is that the Mach 1 region moves to the smallest restriction of the venturi called the *throat*. What happens after the throat can be somewhat complicated but in general it depends on the pressure downstream of the throat. If the pressure is the same as in the tube before the venturi, the air returns to the initial conditions of velocity, pressure, temperature, and density. If the pressure is lower, as in expansion into a large low-pressure volume, the air will expand, causing the velocity to continue to accelerate beyond Mach 1, as shown in Figure 8.17. This further expansion and acceleration results in a rapid decrease in air temperature, density, and static pressure.

> The most powerful rocket made was the Saturn V, which launched the Apollo missions.

This expansion and acceleration rather than deceleration is exactly what happens in the first part of the wing in transonic flight. In subsonic flight the air decelerates and the pressure increases after the point of greatest curvature on the wing. In transonic flight the air accelerates and the pressure continues

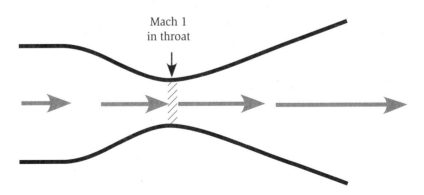

Fig. 8.17. The supersonic venturi.

to decrease after the greatest curvature until the normal shock wave is reached.

ROCKET MOTORS

It should be noted that this description of a supersonic venturi is the same as the description of a rocket motor. The flow of the compressed gas from the combustion chamber is restricted by the region of Mach 1 that has moved to the throat of the motor. The gas then expands into a region of lower pressure so it accelerates. Thus the exhaust of a rocket motor is supersonic. Since by Newton's second law the thrust of the rocket is proportional to the velocity of the gas, this is a very desirable situation.

The velocity of gas is Mach 1 at the throat of all rocket motors. This limits the amount of gas that can be expelled. In designing a motor the engineer must make the throat small enough so that the gas reaches Mach 1 and thus the exhaust becomes supersonic. But the throat must also be large enough to let enough gas out to produce the desired thrust.

Supersonic Wind Tunnels

Supersonic wind tunnels operate differently than subsonic and transonic wind tunnels. First, because fans are inefficient at supersonic speeds, they must run subsonic and the air must make a transition from subsonic to supersonic speeds. Second, supersonic wind tunnels require an enormous amount of power. Supersonic wind tunnels can require so much power that if run during periods of peak electricity demands they can cause a regional *brown-out*. Very few facilities have continuous supersonic wind tunnels for this reason.

The key to making a supersonic wind tunnel is to employ a supersonic venturi. Figure 8.18 shows a schematic of a closed-circuit supersonic wind tunnel. The fan moves the air in a *subsonic channel*. During startup the subsonic section has been pressurized while the test section remains at a static pressure of 1 atmosphere. The air accelerates in the first venturi until the speed at the throat becomes Mach 1. As the channel opens up, since the air is flowing

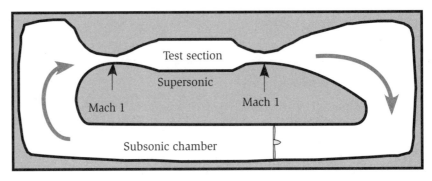

Fig. 8.18. The supersonic wind tunnel.

into a region of lower pressure it accelerates, producing the supersonic flow in the test section. After the test section the airflow goes through a second venturi. Here the speed decreases until it becomes Mach 1 at the throat. Since the air is going into a region of higher pressure, as the channel opens up the flow slows down, becoming subsonic again.

The supersonic wind tunnel has an additional source of power loss. In addition to the friction on the walls and the drag on the models, now there are losses associated with the inevitable shock waves. All of these losses mean a lot of heat is being generated. In order to run continuously, a supersonic wind tunnel must have a large cooler, which is placed in the airflow in the subsonic section.

> The Boeing 747-400 has 171 mi (274 km) of wiring.

The great amount of power required for supersonic wind tunnels means there are very few continuous wind tunnels and they are not very large. A 3 × 3 foot (1 × 1 m) test section is considered very large and requires half a million horsepower (375 megawatts) to operate at Mach 3. But there are other methods to test supersonic aircraft.

One method is the "blowdown" supersonic wind tunnel depicted in Figure 8.19. A huge tank is filled with high-pressure air and then exhausted through a venturi. This kind of wind tunnel works quite well but will only allow a few minutes of testing. However, a carefully planned test can gather a tremendous amount of data in a very short time. With this technique the energy required is generated and stored over time. This type of wind tunnel requires very little power but requires quite a long time between tests. The NASA Hypersonic Tunnel Facility at Plum Brook can generate speed up to Mach 7. This blowdown facility can accommodate a 5-minute test every 24 hours. The Twenty-Inch Supersonic Wind Tunnel at the Langley Research Center can generate flows with Mach numbers from 1.4 to 5 for 1.5 to 5 minutes.

Another option, which is more common, is the *vacuum* supersonic wind tunnel shown schematically in Figure 8.20. Rather than pump a chamber to a high pressure, which is dangerous, the chamber is evacuated and the airflow is in the other direction through the test section. Thus, the upstream reservoir of air is just the atmosphere and the air is being drawn through the throat and test section into a vacuum.

In all supersonic venturis, the air expands on the high-speed side and thus cools. For continuous supersonic wind tunnels this is not a

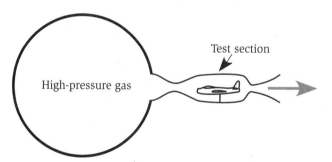

Fig. 8.19. Blowdown supersonic wind tunnel.

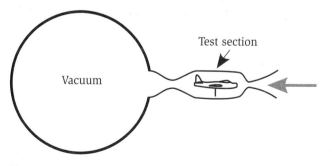

Fig. 8.20. Vacuum supersonic wind tunnel.

concern because all the energy losses cause the air to be hot to start with. For the blowdown wind tunnels the air is often heated before it reaches the venturi so that the test section remains at a reasonable temperature. Vacuum wind tunnels have a problem that the room air is used and thus it is not practical to preheat the air. Therefore, the test section is very cold. For example, a Mach 3 test section would be $-274°F$ ($-170°C$) if the air supply were at room temperature.

Hypersonic Testing

With the incredible power required for supersonic wind tunnels, how can anyone expect to create hypersonic flow conditions, typically above a Mach 5, in a test environment? The only effective method to do this with a stationary model is with the blowdown method, lots of preheating of the air, and a very small test section. The key word in that last sentence was stationary. Some hypersonic facilities actually use a combustion gun, where gases combust in the breach to propel the model. The problem with this technique is that the desired measurements must be made on a nonstationary model, one that is moving very fast.

But there is another trick up an engineer's sleeve. Hypersonic flight implies that the Mach number is typically greater than Mach 5. Up to this point we implicitly assumed that to achieve hypersonic speeds we have to increase the speed in the test section or of the model. What if we were to decrease the speed of sound instead? Sound speed differs for different gases. The speed of sound decreases as the weight of the gas molecules increases. So, instead of using air for our working gas, we could look for a heavier gas, like carbon dioxide, although this will only decrease the sound speed by 14 percent. The advantage of using an alternate gas is that the true speeds can be kept reasonable, while the Mach number is fairly high.

Flight Testing

In this section we explain flight-testing techniques used to verify airplane performance as related to the concepts described previously.

First, it must be understood that flight testing means two very different things to commercial airplane manufacturers and to military airplane manufacturers. To commercial airplane manufacturers, flight testing focuses on meeting FAA requirements. Frequently, an FAA representative will be on board the airplane to monitor the test results. These tests are performed to verify compliance to specific regulations. Because of the expenses, flight tests of commercial transports are rarely performed unless a regulation is involved.

To the military, flight tests usually mean compliance with military specifications. This typically means verifying performance. Because military aircraft fly close to the edge of their operating capabilities, flight testing is used to probe the limits of the airplane. Below we discuss some of the measurements performed in a flight test.

Neil Armstrong was an X-15 (Figure 5.16) test pilot before he entered the astronaut corps.

Flight Instrument Calibration

One of the first steps in flight testing an aircraft is to ensure the altimeter and the airspeed indicator are giving proper information. In most airplanes the altimeter and airspeed indicator use pressure to determine altitude or speed. We start with a little explanation of how they work and then discuss what is tested early in a flight test program.

An altimeter is nothing more than a simple pressure gauge or barometer. The static pressure is measured at the static port, as discussed in the first chapter. The static port is placed somewhere on the surface of the airplane. Because of the airflow around the airplane, the surfaces see a variety of static pressures that are different from the true ambient pressure. But remember, just because the air is flowing faster somewhere it does not mean that the static pressure is lower. But since any place on the airplane is going to see at least some small pressure difference from the ambient static pressure, it is important to calibrate the altimeter.

The goal of the altimeter is to measure the pressure of the atmosphere surrounding the airplane with as little airplane interference as possible. Typically static ports are placed on the fuselage away from the wing, where static pressure changes are smallest. Because the static pressure changes with position on an airplane, the error associated with

Fig. 8.21. Trailing cone on 737-700. (*Used with the permission of the Boeing Management Company.*)

placement of the static port is called *position error*. A flight test must be performed to determine the amount of the position error. To do this the pressure must be measured free from aircraft interference. One method used by commercial airplane manufacturers is to use a *drag cone* to pull a probe far to the rear of the airplane, as shown in Figure 8.21. The drag cone is deployed and then retracted with a winch in the tail of the airplane.

On many military tests, flight testing for position error is measured on a probe mounted on a long *spike* on the nose of the airplane as shown in Figure 8.22. The difference between the true air pressure and that measured by the static port can be translated into an altimeter error, which must be published for the airplane.

In 1955, "Tex" Johnston became a legend at Boeing when he rolled the Boeing 707 prototype at an air show in front of many airline executives.

Fig. 8.22. Airspeed calibration probe on D-558-II. (*Photo courtesy of NASA.*)

THE STANDARD DAY

The pressure variation with altitude is not exactly the same every day, so a "standard day" has been defined and used internationally. A standard day is 59°F and 29.92 inches of mercury (15°C and 1013 millibars) at sea level. The pilot adjusts the altimeter for the true barometric pressure so that correct altitude is indicated. When calling in for permission to land at an airport, among the first things given to the pilot are the barometric pressure and the wind's speed and direction. Below 18,000 feet all altimeters are adjusted to local atmospheric conditions so that two airplanes in the same airspace will both read correctly. Above 18,000 feet, all altimeters are set for the barometric pressure of a standard day.

As discussed in the first chapter, the airspeed indicator works off the difference between the static pressure and the dynamic pressure measured by the Pitot tube. The static pressure is usually taken from the same static port used by the altimeter. Unlike the static port, the value measured by the Pitot tube does not change with position unless the aircraft is flying supersonic. But since the static port is position-sensitive, position error also affects the accuracy of the airspeed indicator.

The German aeronautical pioneer Otto Lilienthal died after his glider stalled and crashed. As he lay dying, he was quoted as saying "sacrifices must be made."

Power Required

Once the altimeter and airspeed indicator are calibrated we can use these, along with engine-power calculations for piston-driven airplanes or thrust for jets, to determine the power and drag curves for straight-and-level flight. In principle one can get both curves and determine the induced and parasitic powers and drags by measuring the power at only two speeds.

POWER REQUIRED DATA

Though a little beyond the scope of this book, if one plots power times speed as a function of speed to the fourth power, the result is a straight line. This is illustrated in Figure 8.23 taken from the flight manual for a Cessna 172. Although values for four points were used, in principle the same results could be obtained from only two points. From this plot one can produce a plot of power as a function of speed (the power curve) and also a plot of drag (power/speed) as a function of speed (the drag curve). Knowing how induced power and drag go as speed, one is also able to separate out the different components of power and drag. This is a lot of information from two simple measurements.

The above measurements were made at one weight and altitude. But we know that both weight and air density affect the power required for straight-and-level flight. To perform the test described above for different altitudes and different loads would be expensive and time-consuming. Engineers have developed the *equivalent weight system* to avoid these additional tests. The air density at the test altitude is determined from pressure and temperature measurements and related to sea-level density. Similarly, the weight of the airplane during the test is measured and related to a standard weight, usually either the maximum gross weight or the empty weight. Known relationships already discussed for power requirements as a function of weight and air density allow the data to be related to any altitude and weight regardless of actual altitude and weight flown on the flight test.

The first U.S. jet fighter was the Lockheed P80. It was conceived in 1943 to counter the German ME-262. The P80 scored the first aerial combat victory between two jet fighters when one downed a MIG-15 over Korea on November 8, 1950.

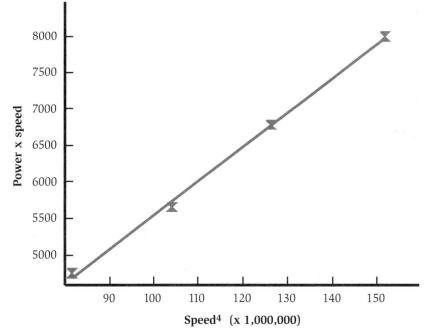

Fig. 8.23. Flight data of power times speed as a function of speed to the fourth power for a Cessna 172.

Takeoff and Landing

Flight testing of takeoff performance is one of the more extensive tests required for commercial airplanes. Certain aspects, such as friction with the ground, depend on runway conditions such as the presence of ice or water. The designer made assumptions about friction, which have to be verified in flight test. Once a few conditions are verified, usually calculations are used to fill in the rest of the operating procedures.

Takeoff techniques also have to be established. For example, at what speed should a pilot begin rotation? At what speed should the airplane lift off? If the speed is too low, the airplane might be in ground effect on the backside of the power curve and unable to climb.

Two tests are particularly exciting. They are the VMU (velocity-minimum-unstuck) test and the maximum braking test. VMU is the minimum speed at which the airplane can leave the ground (that is "unstuck"). This occurs at approximately the stall angle of attack. In order to achieve this goal, part of the airplane tail may actually drag on the ground. When aluminum hits concrete, sparks tend to fly. It

Fig. 8.24. VMU test with a tail skid on the airplane to prevent damage. (*Used with the permission of the Boeing Management Company.*)

can look as if the airplane is on fire, as shown in Figure 8.24. Usually little damage is done to the airplane. For flight testing, a small tail skid is placed on the airplane to prevent damage.

The maximum-braking test demonstrates that the airplane can abort takeoff and stop on the runway without risking the passengers. What makes this test exciting is that all of the kinetic energy of the airplane is transferred to the brakes. So the brakes get extremely hot, as discussed in the previous chapter. The test requires that the airplane be able to remain on the runway without help for a certain period of time. You have to understand that the tires are probably melting and exploding. The heat from the breaks may be radiating to the underside of the fuel tanks. The test ensures that if a maximum-breaking abort is required, the airplane can survive until fire trucks can arrive on the scene to cool down the brakes.

Takeoff and landing tests must be performed in a variety of configurations. These are demonstrating the "what happens if?" scenarios. Tests with one engine out, flaps in various positions, different gross weights, and at different atmospheric conditions must be performed.

Although the first supersonic flight did not occur until 1947, Ernst Mach photographed the shock waves on a supersonic bullet in 1887.

As computer models get more and more accurate, some flight testing is being replaced by careful calculations. So, rather than having to test every possible condition, which is very expensive, a few key conditions can be tested in flight and used to validate the calculations. The rest of the conditions can then be determined by analysis. The predictive abilities are so good today that there are rarely surprises in performance when the airplane goes for its flight test.

Climbing and Turning

In the previous chapter it was shown that climbs and turns require power and thrust. So climbing and turning tests are nothing more than determining the available power. The difference between the power available and the required power is the excess power. The bottom line is that the power and thrust available is the only test required to determine climb and turn performance.

To measure the available power, the procedure is to push the throttle to a specific setting and measure the acceleration. The acceleration times the mass of the airplane is the net thrust. Since the required power and thrust can be calculated from a previous flight test, the available thrust, and available power, can be determined.

Flight Test Accidents

Unfortunately, flight test accidents do not leave much to talk about. Today, there is rarely a crash of a commercial airplane in flight testing. With military aircraft it happens on rare occasions, but nothing like during the rapid development period of the 1940s to 1960s. Today even military aircraft performance is so well predicted that there are few surprises. But, in general, the riskier business of military flying will lead to more flight-test accidents. The bottom line is that flight testing occurs all the time with very few accidents.

Lt. Thomas Selfridge was the first casualty of a powered airplane crash. He was assigned by the U.S. Army to ride with Orville Wright. A cable snapped, breaking a propeller, leading to the crash.

Wrapping It Up

Testing is an integral part of any airplane development program. It is an expensive part too. The days are long gone when testing can be used in place of detailed analysis to develop an airplane design. Testing is now used mostly for refinement and validation.

This book has taken you through a complete course in aeronautics. We hope that you have achieved an intuitive understanding of flight.

It is likely that you understand flight on a more profound level than what is often achieved with courses that use much more mathematics and much less physics.

In one way we have not done you a favor. You will likely now be sensitized to the mythology that is commonly expounded in books, on TV, and on the Web. We hope that you will find compensation in a better understanding and wonder about the marvels of modern aviation.

Misapplications of Bernoulli's Principle

ernoulli's equation has mistakenly become linked to the concept of flight. "Demonstrations" of Bernoulli's principle are often given as demonstrations of the physics of lift. They are truly demonstrations of lift but certainly not of Bernoulli's principle. As discussed in Chapter 1, we are often taught only part of what is necessary to understand the applications of Bernoulli's equation. This has been the source of a great many misconceptions that have been enthusiastically propagated. When we are first introduced to Bernoulli's equation it is always in respect to a fluid flowing in a pipe with a restriction. Since mass must be conserved, the flow through the smaller cross section is faster. Since energy is conserved, the faster-flowing fluid has a lower pressure. This is often all that we are taught. And from this, most of us have come away with the belief that if air is moving faster it has a lower pressure. On giving it deeper thought we might rightfully assume that the lowered pressure is measured perpendicular to the flow (the *static pressure*), since we know that if we put our hand in the path of this faster-flowing air we would feel an increase in pressure.

Because of this point of view, some very interesting things are taught. First, take the example of the "Bernoulli strip" illustrated in

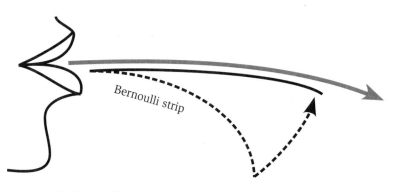

Fig. A.1. The Bernoulli strip.

Figure A.1. The Bernoulli strip is a narrow piece of paper which one blows across the top to produce lift. It rises into the airstream and is clearly an example of lift on a wing. "The air goes faster over the top, thus the pressure is lower and the paper rises." Or at least so goes the explanation.

Another common example of the physics of lift is that of a Ping-Pong ball supported by a vertical jet of air (Figure A.2a). The argument is that since the air is moving the static pressure is lower. When the ball moves to the side, it comes into contact with the still air that is of a higher pressure. The ball is pushed back into the flow.

Before we explain what is missing in our understanding of Bernoulli's education, let us revisit the *static port* on an airplane. An airplane has a small port somewhere on its side of the fuselage where the static pressure is measured by the instruments, such as the altimeter. This port provides a fairly accurate static pressure measurement, even though air passes over it at a high speed. If one watches the altimeter when the engine is started and the propeller blows air across the static port, the indicated altitude does not change. But the altimeter gives a very sensitive measure of pressure. So what is wrong with our understanding of the Bernoulli principle?

In aeronautics, Bernoulli's equation is well understood. Ignoring the change in altitude and compressibility of an airflow, one can write Bernoulli's equation as:

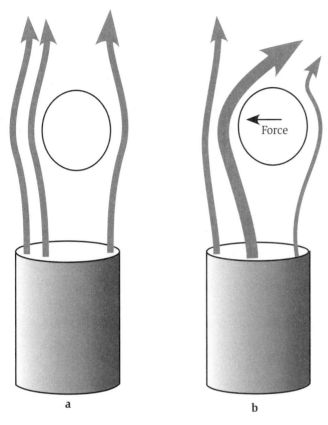

Fig. A.2. Ping-Pong ball in a jet of air.

$$P_{static} + \tfrac{1}{2}\,\rho\,v^2 = P_{total}$$

Here P_{static} is the static pressure measured perpendicular to the flow. The second term is referred to as the *dynamic pressure* where ρ is the density of the air and v is its speed. Thus, the dynamic pressure is a measure of the kinetic energy of the air. P_{total} is the *total pressure*. All this was discussed in Chapter 1.

In a confined pipe the sum of the dynamic pressure and the static pressure is a constant. If we know that constant, the static pressure can be calculated from a knowledge of the air's velocity. The same is true for air that is given energy, say by a propeller or our breath. But now we do not know the constant any more. In fact, kinetic energy has been given to the air. Thus the dynamic pressure has increased, but the static pressure has not decreased. The fact that the air is moving faster does not necessarily mean that the static pressure has decreased.

Let us now look again at the Ping-Pong ball in the jet of air. First, one might reason that since the jet of air is not confined, if it had a lower static pressure, the surrounding air would collapse the jet until it had the same static pressure as the surroundings. This is reasonable, since there would be a difference in (static) pressures and no barrier to separate them. In fact the source of the jet of air has only increased the dynamic and total pressures of the air. Likewise, one's breath does not have a decreased static pressure. Thus one must look for another explanation for the Ping-Pong ball swinging together and the lifting of the Bernoulli strip.

The answer lies in the Coanda effect and Newton's laws discussed in Chapter 2. Remember, the Coanda effect is the pheromone that causes a flowing fluid such as air to wrap around a solid object. When the ball is near the edge of the jet of air, the Coanda effect causes an asymmetric flow of air around the ball, as in Figure A.2b and momentum transfer causes a force to push the ball back in, just like the lift on a wing.

The same is true with the Bernoulli strip. The Coanda effect causes the air to bend over the paper strip. Newton's first law says that this requires a force on the air. Newton's third law says that an equal and opposite force is exerted on the paper. The paper is lifted. Our incomplete understanding of its application causes most of the problems with the applications of Bernoulli's principle. We have been led to assume that if air is flowing its static pressure has been lowered. This of course is not necessarily so.

There are two other phenomena often attributed to Bernoulli's principle. The first is the situation where one blows between two Ping-Pong balls hanging on strings as shown in Figure A.3. The result is that they swing in toward each other. Here we just have the same phenomenon as the Ping-Pong ball in the jet of air. But in this case there are two balls instead of one.

A more interesting misapplication of Bernoulli's principle is in the explanation of the curve flight of a spinning baseball. Let us start the discussion by examining the airflow around a nonspinning ball in flight, as shown in Figure A.4a. In the figure the ball is

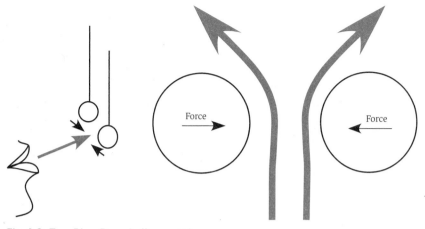

Fig. A.3 Two Ping-Pong balls on strings.

traveling from right to left. The air splits around the ball and the only force on the ball is the drag of the air.

Figure A.4b shows the effect on the air when the ball is spinning. Here we have removed the effect of the ball traveling through the air. The roughness of the ball causes air to be dragged around forming a boundary layer moving in the direction of rotation. The stitching on the ball enhances this boundary layer formation. The surface is sometimes illegally roughened by the pitcher to enhance the effect.

When we put the airflows of Figure A.4a and A.4b together we find that the airflow looks very much like that over a wing only upside down in this example. The air that travels over the op of the ball meets the oncoming flow around the ball and loses energy. This causes the air to separate from the ball fairly early. The air that goes under the ball is traveling in the same direction as the air around the ball. This air does not lose energy as the air passing over the top of the ball did and thus stays attached longer. The result of all this is that there is a net upwash behind the ball and thus a downward force on the ball. So a spinning ball will feel a sideways force when traveling through the air. The backspin given to a dimpled golf ball causes it to experience a lifting force in the same way.

There are other examples of the misapplication of Bernoulli's principle, but we think you get the idea. The next time you hear Bernoulli given credit for some phenomenon, think it through and see if you really believe what you are being told.

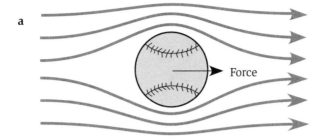

a

Force

b

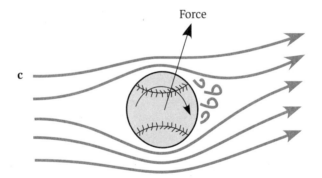

Force

c

Fig. A.4. The curve ball.

Index

About the Authors

David F. Anderson is a physicist at the Fermi National Accelerator Laboratory and a private pilot.

Scott Eberhardt is an associate professor in the department of aeronautics and astronautics at the University of Washington. He is also the director of the Kirsten Wind Tunnel and a private pilot.